Geothermal Energy - Challenges and Improvements

Edited by Zayre Ivonne González Acevedo and Marco Antonio García Zarate

Published in London, United Kingdom

Geothermal Energy - Challenges and Improvements
http://dx.doi.org/10.5772/intechopen.102297
Edited by Zayre Ivonne González Acevedo and Marco Antonio García Zarate

Contributors
Héctor Aviña-Jiménez, Eduardo Pérez-González, Rodrigo Alarcón-Flores, Lilian Namuma S. Kong'ani, Raphael M. Mulaha Kweyu, Eléonore Dalmais, Albert Genter, Guillaume Ravier, Vincent Maurer, David Fries, Béatrice Pandélis, Helmut Duerrast, Zayre Ivonne González Acevedo, Marco Antonio García Zarate

First published in London, United Kingdom, 2023 by IntechOpen
IntechOpen is the global imprint of INTECHOPEN LIMITED, registered in England and Wales, registration number: 11086078, 5 Princes Gate Court, London, SW7 2QJ, United Kingdom

British Library Cataloguing-in-Publication Data
A catalogue record for this book is available from the British Library

Additional hard and PDF copies can be obtained from orders@intechopen.com

Geothermal Energy - Challenges and Improvements
Edited by Zayre Ivonne González Acevedo and Marco Antonio García Zarate
p. cm.
Print ISBN 978-1-80356-998-7
Online ISBN 978-1-80356-999-4
eBook (PDF) ISBN 978-1-83880-282-0

Meet the editors

Zayre I. González Acevedo obtained a bachelor's degree in chemical engineering, specializing in the environment, from the Technological Institute of Toluca, Mexico, in 1998; a master's degree in chemical engineering (process integration) from the University of Guanajuato, Mexico in 2002; and a Dr. rer. nat. (*magna cum laude*) in environmental geochemistry from the University of Heidelberg, Germany, in 2006. She was a researcher at Mexico's National Institute of Nuclear Research from 2006 to 2012 and has been at the Center for Scientific Research and Higher Education at Ensenada, Baja California, since 2012. She was responsible for Work Package 9, 'Environmental, Social and Economic Impacts of Enhanced and Super-Hot Geothermal Systems' in the GEMex project 'International Cooperation in Research and Development between Mexico and the European Union in Geothermal Energy' (2018-2021), and Project 25, CeMIE-Geo, 'Sustainable Development and Environmental Impact Assessment of three Geothermal Exploration Zones with Exploitation Potential in Mexico' (2014-2018).

Marco Antonio García Zarate specialized in the Optical Laboratory of the Applied Physics Department of the CICESE(Center for Scientific Research and Higher Education of Ensenada) with a CONACYT (National Council of Science and Technology in Mexico) scholarship from July 1980 to July 1981. He received a degree in Industrial Engineering from the Technological Institute of Ensenada in 2010. Later, Dr. Zarate pursued a Master of Science in Arid Zone Ecosystem Management at the Faculty of Science of the Autonomous University of Baja California, with a CONACYT scholarship, and received Honourable Mention in August 2013. He received his Ph.D. in Environment and Development from the Autonomous University of Baja California, Mexico in October 2016., with a CONACYT scholarship and Honourable Mention. In February 2017, he was awarded the Scholar Merit by the Autonomous University of Baja California. Currently, Dr. Zarate is a teacher at the undergraduate, master's, and industrial levels, and has been working at the Centre for Scientific Research and Higher Education in Ensenada, Baja California since 1997.

Contents

Preface **IX**

Section 1
Introduction **1**

Chapter 1 **3**
Introductory Chapter: Geothermal Energy – Challenges and Improvements
by Zayre Ivonne González Acevedo and Marco Antonio García Zarate

Section 2
Challenges and Improvements **9**

Chapter 2 **11**
Geothermal Energy for Southern Thailand: Opportunities and Realities
by Helmut Duerrast

Chapter 3 **29**
Toward Sustainable Implementation of Geothermal Energy Projects – The Case of Olkaria IV Project in Kenya
by Lilian Namuma S. Kong'ani and Raphael M. Kweyu

Chapter 4 **47**
Environmental and Socio-Economic Impact of Deep Geothermal Energy, an Upper Rhine Graben Perspective
by Eléonore Dalmais, Guillaume Ravier, Vincent Maurer, David Fries, Albert Genter and Béatrice Pandélis

Chapter 5 **73**
General Information on Geothermal Energy
by Héctor Aviña-Jiménez, Eduardo Pérez-González and Rodrigo Alarcón-Flores

Preface

Geothermal energy is the heat that comes from the inner layers of the Earth. This heat can be used directly in the production processes such as heating buildings, aquaculture, farming, and so on, or indirectly to produce electricity from the water steam that comes from deep (1000–3000 m) geothermal wells. The source of geothermal energy is mainly attributed to the natural movement of oceanic and continental tectonic plates, which when colliding or separating from geologically active regions in which the phenomena of plate subduction or divergent plates occur causes hydrothermalism. Therefore, a geothermal system is made up of three main elements: a heat source, a reservoir, and a fluid, which is the medium that transfers the heat. The heat source can be either a magmatic intrusion at a very high temperature (> 600°C) or a reservoir.

Within the framework of the fight against climate change, geothermal energy is becoming increasingly important among renewable energy sources due to its mature technology, efficiency, and sustainability. In addition, its gas emissions are mainly water steam, reducing the negative impact on our planet. However, only 15% of the world's known geothermal reserves are exploited for global electricity production, which constitutes a very small fraction of the immense amount of energy available on Earth. Today, many countries have taken advantage of their geothermal resources. The United States and the Philippines are the largest producers of electricity from geothermal energy.

A geothermal power plant has several advantages, including reliability, independence and the fact that it is relatively inexhaustible, emits few pollutants, uses smaller land compared with solar energy, and contributes to the development of rural areas.

Recent scientific studies related to the development of improved exploration and exploitation techniques for new generations of geothermal systems show that, in the medium term, geothermal power generation will become a key element in the energy mix. Not only as a commitment to the use of renewable energies, but also as a commitment to technological development and the creation of economic activity and, therefore, employment.

Geothermal energy is an energy source with great potential. The installation of renewable energies can lead to greater performance, efficiency, and results. As such, this book provides four representative case studies from Africa, Asia, Europe, and America, highlighting the social, economic, and environmental challenges these countries have addressed to developing geothermal energy and using geothermal heat to ensure the sustainability of the resource. It also discusses the different aspects of geothermal energy, including its social repercussions and effects on the environment, as well as public policies and management for better regulation of planning and environmental protection.

Zayre Ivonne González Acevedo
Geology Department,
Center for Scientific Research and Higher Education at Ensenada (CICESE),
Baja California, Mexico

Marco Antonio García Zarate
Applied Physics Department,
Center for Scientific Research and Higher Education at Ensenada (CICESE),
Baja California, Mexico

Section 1

Introduction

Chapter 1

Introductory Chapter: Geothermal Energy – Challenges and Improvements

Zayre Ivonne González Acevedo and Marco Antonio García Zarate

1. Introduction

1.1 Outline

In recent years, the issue of energy production has been in the international debate. The discussion has focused on the generation of agreements and actions for the care of the environment, due to the negative impacts that the energy industry, mainly based on fossil fuels, has generated on the planet [1, 2].

The use of natural resources in a more responsible and comprehensive manner has become more relevant. The energy crisis and climate change have pointed to the development of technologies that allow the use of alternative energies with higher performance, efficiency, and less environmental impact. Hence, geothermal energy plays an important role to be used as an alternative energy source, which contributes to the achievement of the Sustainable Development Goals [3]. For example, SDG 7, on affordable, reliable, sustainable, and modern energy for all, SDG 13, on climate action and the Paris Agreement [4]. Being recognized as an ally to solve part of the problem of climate change and distribute electricity at low economic, social, and environmental cost [5].

Among the alternative energies, geothermal energy is the heat energy generated in the interior of the Earth. Under favorable conditions, a small proportion of this energy can be extracted and used by humanity. A geothermal system is the combination of the following elements: a heat source, a fluid that transfers heat, porous and permeable rocks that allow the accumulation of the hot fluid, and impermeable rocks that function as a sealing layer to prevent or reduce the migration of the hot fluid to the surface. The deeper it is the fluid, the higher is its temperature, and according to its enthalpy, it has the capacity to provide direct uses or to generate electricity [6–8].

Geothermal resources have been identified in almost 90 countries, with a record of geothermal utilization in more than 70 countries. As of 2010, electricity from geothermal energy is produced in 24 countries. Nearly 40 countries worldwide are considered to have sufficient geothermal potential that could meet their total electricity demand with geothermal energy, for example, Costa Rica, Ecuador, Guatemala, Indonesia, Iceland, Mozambique, Peru, and the Philippines. While Iceland and El Salvador have the highest share of geothermal energy in their country's energy mix, generating about 25% of their electricity from geothermal resources. The United States and the Philippines have the largest installed capacity of geothermal power plants: approximately 3000 MW and 1900 MW, respectively [9].

IntechOpen

2. Advantages

The great advantage of the geothermal resource is that its transformation into electricity can be done independently of the weather and a schedule. As is the case of wind and solar energy, playing an important role since it is available 24 hours a day, 365 days a year, allowing continuously feed to the geothermal power plants, which can act as the economic base of a region [10]. In addition to requiring, less land per megawatt produced, compared with solar and wind energy.

This energy resource has many advantages over other alternative energies when Life Cycle is considered in the analysis, such as competitive heat prices, continuous source of energy, and low environmental impacts. Furthermore, the additional valorization of geothermal water through its use for low-temperature heating and the recovery of mineral resources are ways to provide additional benefits to the local communities and/or the developer, in a sustainable environment in the exploitation of the direct and indirect uses of the geothermal resource [11]. To achieve this, a clear energy policy is needed in the countries, between the developer, government, and communities. The lack of commitment and enthusiasm of governments weakens the potential growth of the sector to be developed by private sector investment or foreign investors, to activate the production and improve the productivity of geothermal energy [12].

3. Challenges

One of the challenges of geothermal energy is that it is closely linked to the geological structures present in active volcanic areas, in areas with geological faults and areas with seismic activity, which represents the uniqueness of the resource. Its extraction is a challenge, which signifies the need to improve exploration and exploitation techniques to access deep resources and avoid negative impacts on the environment. However, due to its nature, it has several opportunities for technological development and productive processes for human development with direct uses of geothermal heat or geothermal power plants [13].

Despite that, geothermal energy is an alternative source of electricity, it is sometimes considered as a renewable energy (when extraction does not exceed recharge) [14], a sustainable energy (when production rates are maintained for more than 40 years) [15], and a clean energy (when CO_2 concentrations emitted are below the local limits) [16]. However, it can cause negative impacts on the environment, which sometimes trigger social conflicts.

Nowadays, it is known that environmental conflicts caused by the execution of geothermal projects introduce new variables between the developer, the government, and the communities [17]. Therefore, it is important to solve conflicts related to this natural resource. Through mediation between key actors, or by proposing management strategies for geothermal projects, which need a high level of acceptance by the inhabitants of the surrounding area, being the local factor crucial to minimize the socioeconomic impact and promote social acceptance.

Author details

Zayre Ivonne González Acevedo* and Marco Antonio García Zarate
Center for Scientific Research and Higher Education at Ensenada (CICESE), Baja California, Mexico

*Address all correspondence to: zgonzale@cicese.mx

References

[1] Martins F, Felgueiras C, Smitkova M, Caetano N. Analysis of fossil fuel energy consumption and environmental impacts in European countries. Energies. 2019;**12**(6):964

[2] Solarin SA. An environmental impact assessment of fossil fuel subsidies in emerging and developing economies. Environmental Impact Assessment Review. 2020;**85**:106443

[3] Majid MA. Renewable energy for sustainable development in India: Current status, future prospects, challenges, employment, and investment opportunities. Energy, Sustainability and Society. 2020;**10**(1):1-36

[4] Sachs JD, Schmidt-Traub G, Mazzucato M, Messner D, Nakicenovic N, Rockström J. Six transformations to achieve the sustainable development goals. Nature Sustainability. 2019;**2**(9):805-814

[5] Byrne JA, Lo AY, Jianjun Y. Residents' understanding of the role of green infrastructure for climate change adaptation in Hangzhou, China. Landscape and Urban Planning. 2015;**138**:132-143

[6] Barbier E. Geothermal energy technology and current status: An overview. Renewable and Sustainable Energy Reviews. 2002;**6**:3-65

[7] Bruni S. La Energía Geotérmica en una nueva serie sobre la innovación de energía. Centro de Innovación Energética (CIE). Washington DC, USA: Departamento de Infraestructuras y el Medio Ambiente, Banco Interamericano de Desarrollo (BID); 2014. pp. 1-10

[8] Nicholson K. Geothermal Fluids: Chemistry and Exploration Techniques. Springer Science & Business Media; 2012

[9] Gehringer M, Loksha V. Geothermal Handbook: Planning and Financing Power Generation. ESMAP Technical Report 002/12. Washington DC, USA: World Bank; 2012

[10] Tomaszewska B, Pająk L, Bundschuh J, Bujakowski W. Low-enthalpy geothermal energy as a source of energy and integrated freshwater production in inland areas: Technological and economic feasibility. Desalination. 2018;**435**:35-44

[11] Shortall R, Davidsdottir B, Axelsson G. Geothermal energy for sustainable development: A review of sustainability impacts and assessment frameworks. Renewable and Sustainable Energy Reviews. 2015;**44**:391-406

[12] Huttrer GW. Geothermal power generation in the world 2015-2020 update report. In: Proceedings World Geothermal Congress. Vol. 1. 2020. p. 15

[13] Soltani M, Moradi Kashkooli F, Dehghani-Sanij AR, Nokhosteen A, Ahmadi-Joughi A, Gharali K, et al. A comprehensive review of geothermal energy evolution and development. International Journal of Green Energy. 2019;**16**(13):971-1009

[14] Stefansson V. The renewability of geothermal energy. In: Proceedings World Geothermal Congress 2000 Kyushu - Tohoku, Japan. May 28 - June 10, 2000. pp. 883-888

[15] Rybach L. Geothermal energy: Sustainability and the environment. Geothermics. 2003;**32**:463-470

[16] González-Acevedo ZI, García-Zarate MA. Geothermal energy as an alternative to reduce atmospheric emissions and provide green energy.

In: Green Technologies to Improve the Environment on Earth. London, UK: IntechOpen; 2018

[17] Gabo-Ratio JA, Fujimitsu Y. Exploring public engagement and social acceptability of geothermal energy in the Philippines: A case study on the Makiling-Banahaw geothermal complex. Geothermics. 2020;**85**:101774

Section 2

Challenges and Improvements

Chapter 2

Geothermal Energy for Southern Thailand: Opportunities and Realities

Helmut Duerrast

Abstract

Electrical energy demand for Southern Thailand is continuously increasing, with new coal/gas-fired power plants planned. However, coal/gas-fired power plants are not only large CO_2 emitters, thus intensifying the on-going climate change crisis, but also their technology costs remain stagnant at comparable high levels. Solar and wind energy can be produced at far lower costs; however, their shares on the renewable energy mix are comparably small in Thailand, but with steady increase. A disadvantage of solar and wind energy is that the production is not constant due to day/night and weather, respectively. Such can be compensated by adding geothermal energy, which can act as a backbone of the renewable energy mix, although absolute amounts might be relatively low. In Southern Thailand, hot springs are the surface expressions of active geothermal systems at depth. Surface exit temperatures can reach up to 80°C and reservoir temperatures up to 143 °C, thus being considered as low enthalpy resources, which can be utilized applying binary power plant technology. In the current renewable power plant, geothermal energy is not considered, but Southern Thailand holds promising quantities of geothermal resources. The only current geothermal power plant in Thailand located in Fang can act as a positive example.

Keywords: geothermal resources, hot springs, geothermal energy, binary technology, policy

1. Introduction

The current and ongoing climate change is mainly the result of the human-made emissions of carbon dioxide (CO_2), as well as other greenhouse gases, for example, methane (CH_4). At the beginning of the CO_2 concentration measurements in the atmosphere at the Mauna Loa Observatory, Hawai'i, USA, the CO_2 value was around 313 ppm (March 29, 1958) and around 200–300 ppm during the approximate 800,000 years before. Today, the CO_2 concentration in the atmosphere stands at around 420 ppm as of July 2022 [1]. Scientific evidence has clearly shown that the increase in greenhouse gas emissions is already resulting in a warming at global scale and will continue to do so [2], with subsequent changes in rainfall patterns and continental aridity, for example [3]. Recent data show that the first 3 months of 2020

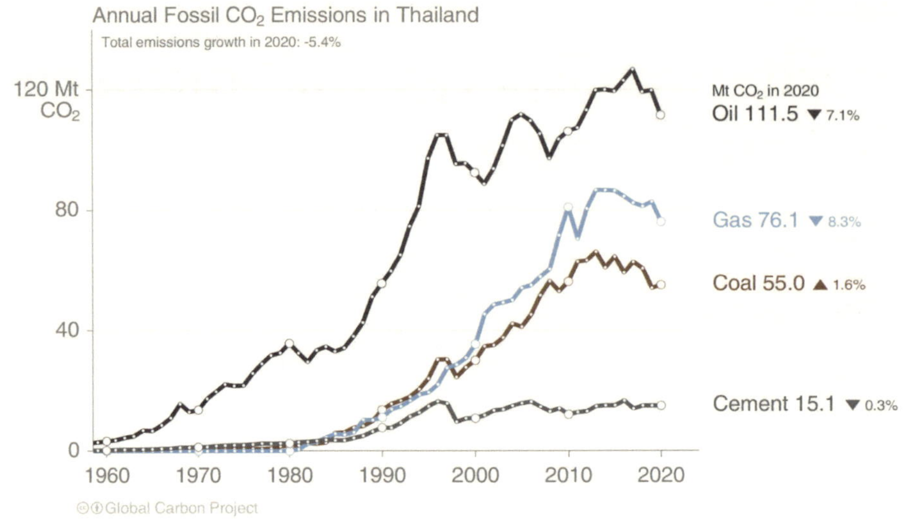

Figure 1.
Annual fossil CO_2 emissions in Thailand in Mt. CO_2 with the growth rates of each sector in percent for 2020; the total emission growth in 2020 was −5.4% due to COVID-19 pandemics (from [5]; used with permission of the global carbon project under the creative commons attribution 4.0 international license).

were the second warmest on meteorological record, only being superseded by the year of 2016 with a strong El Niño observed, for example, [2]. Subsequently, mitigation efforts of the climate change induced global warming require an accelerated decarbonization of all relevant sectors in industry and society at global scale, from energy, over transport, to heating/cooling, in order to reduce significantly the CO_2 emissions by human activities and thus meet the Paris Climate Agreement with a 1.5°C goal [4].

CO_2 emissions from burning fossil energy resources and other sources have increased steadily over the last 70 years and even after a dip down during the height of the COVID-19 pandemics in 2020, CO_2 values are project to bounce back for 2021 [5]. For 2020 global fossil, CO_2 emissions originated from the following main sources are projected as follows [5]: around 41% from coal/lignite, 32% from oil, 22% from natural gas, 5% from cement production. Other smaller sources include gas flaring during petroleum exploitation, steel, and petrochemical industry as well as refineries. For year 2020, it is further projected that about 61.5% of this CO_2 is emitted by only 10 countries according to [5], with China (24.2%), USA (15.9%), India (4.7%), Russia (4.5%), and Japan (3.4%). Thailand's global CO_2 contribution for the same year stands by 0.72% (global rank #24 from 221 countries/regions). From this 43.3% are from oil, 29.5% from gas, 21.3% from coal, and 5.9% from cement production (see [5]; **Figure 1**).

2. Southern Thailand

For Southern Thailand, the electrical energy supply is currently maintained by mainly conventional gas and diesel-powered units as well as hydro dams and to a minor extent by biomass/gas systems, mostly related to agro-industries, as well as electricity add from central Thailand and to minor extent imports from Malaysia (see [6]; **Figure 2**). The conventional gas and diesel-powered units are as follows:

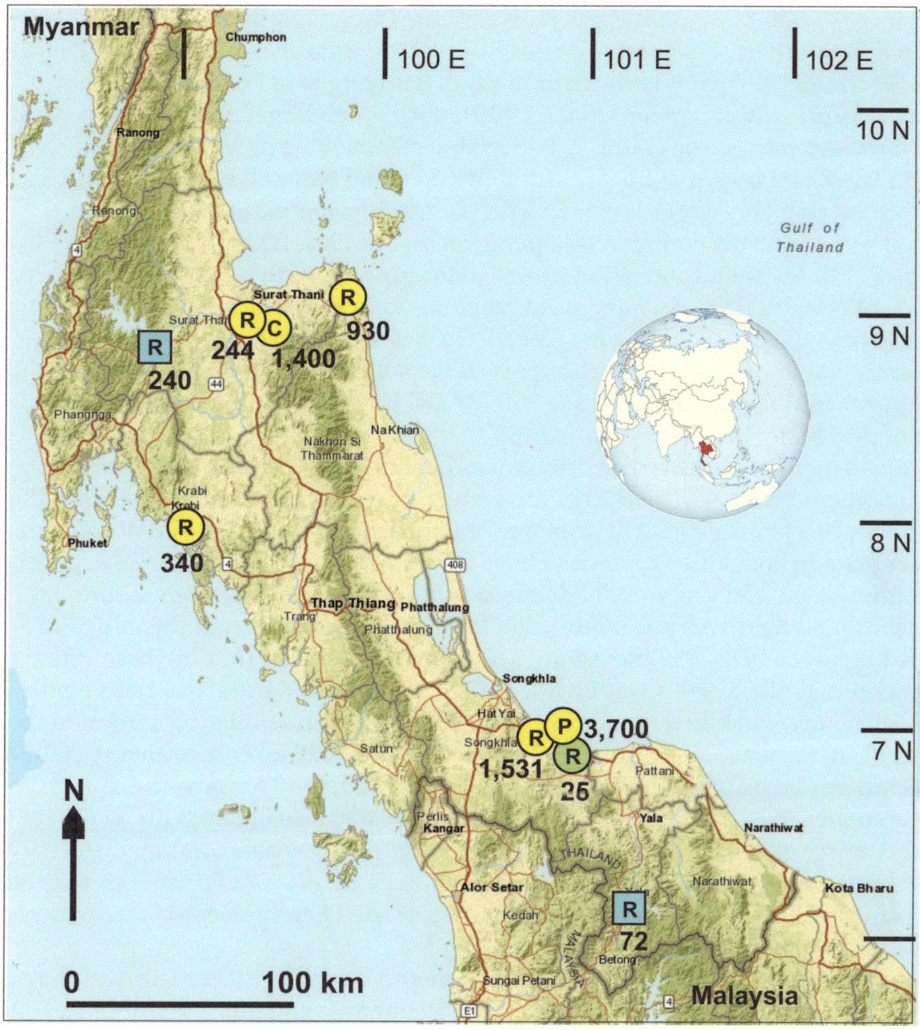

Figure 2.
Electricity generation in Southern Thailand. Status: R = running, operating; C = under construction; P = planned. Fuel source: Yellow = oil, gas; green = biomass; blue-hydro. Power: Values next to symbols, in MW (further information and references in the text). Globe via [7].

(1) In Songkhla's Chana District are combined 1531 MW natural gas-fired power plants, which are connected to the Thailand Malaysia Joint Development Area's (JDA) gas field in the Gulf of Thailand *via* a pipeline. (2) Smaller units with 244 MW natural gas and diesel-powered plants are located in Surat Thani's Phun Phin District. (3) In Krabi's Nuea Khlong District is a 340 MW a fuel oil-powered plant. Here, diesel has to be transported *via* small ships through mangrove areas protected under the Ramsar Convention [8]. The Krabi power plant was originally lignite fired as deposits were found in the area nearby (Krabi Basin, geologically), but mining has finished since more than a decade. Attempts to replace diesel by coal were put on hold and finally discarded as it would require a new port facility to handle the coal as well as several kilometer long conveyor belt systems to transport the coal to the power plant [9]. (4) In Khanom District of Nakhon Si Thammarat

Province is a 930 MW natural gas power plant. Two major hydropower dams are located in Southern Thailand: (5) the 240 MW Rajjaprabha Dam in Surat Thani and (6) the 72 MW Bang Lang Dam in Yala's Bannang Sata District [10]. Both dams were already completed in the 1980s, with much larger dams in the central and northern part of the country. Renewable electricity-producing units are, for example, (7) a 2.062 MW biogas unit in Krabi using palm oil wastewater for methane production [11]. (8) A larger 25 MW biomass power plant using rubber trees as fuel source started commercial operation on March 1, 2020 in Songkhla's Chana district [12]. Further, three wind power plant projects with a combined capacity of 126 MW in Nakhon Si Thammarat and Songkhla province are currently under construction [13]. (9) Additional electricity is channeled *via* 115 kV and 230 kV transmission lines from the central part to the southern region and electricity is also purchased from Malaysia *via* a 300 kV DC line with a maximal transmission rate of 300 MW [14].

Due to increasing electricity demand and in line with the Thailand Power Development Plan 2015 (PDP 2015), the government *via* the Electricity Generating Authority of Thailand (EGAT) proposed the construction of a 2200-MW coal-fired power plant in Songkhla Province (Thepa District) by 2024 [15]. The coal supposed to come *via* new and yet-to-build deep sea ports with shipments from Indonesia, Australia, and South Africa. A few years later, these plans were put on hold and according to the PDP 2018, Revision 1 [16], the electricity gap left by these still proposed coal plants will so far be filled with two units of natural gas-fired power plants having a combined capacity of 1400-MW, which are under construction in Surat Thani Province and coming online in 2027 and 2029, to ensure energy stability for Southern Thailand [17]. However, in January 2020, the Thai government outlined a new special economic zone in Chana district, located also in Songkhla province, but further north of Thepa, and proposed four power plants with a combined electricity-generating capacity of 3700 MW [18]. Fuel sources are not mentioned, but very likely the plants will follow somehow the blue prints of the Thepa power plant as outlined above.

For Southern Thailand, the electricity demand side comprises mainly of the main tourists areas in Phuket, Krabi, and parts of Phang Nga and Surat Thani (e.g., Koh Samui), located near the shore lines of the Andaman Sea and the Gulf of Thailand. There, also the sea food processing and cold storage facilities are located. Further, significant demand is coming from the larger area of Hat Yai, a commercial center for Southern Thailand, and Songkhla, both with seafood processing and rubber processing companies. The situation in the southern part of Thailand reflects the overall situation of the country, where the electricity generation is dominated by gas due to the discovery of mainly gas and less oil fields in the Gulf of Thailand [6, 19]. For the new special economic zone in Chana district, Songkhla Province, a number of industries are proposed, including petrochemical plants and others [18], following somehow the development and blue print of the Eastern Economic Corridor (EEC), which comprises Chachoengsao, Chon Buri, Rayong, Bangkok, and Samut Prakan Province [20].

3. Geothermal resources and energy

Geothermal energy is exploiting the heat inside the Earth as the temperature in general increases with depth; this separates it from the other main renewable energy source, solar photovoltaic (PV) and wind, which both utilize external energy sources. However, the heat flow from the interior of the Earth to the surface is not uniform across the globe, and it is mainly directed by the local and regional tectonic setting, especially in relation to lithospheric plate boundaries [21]. In general, areas along divergent (extensional) plate boundaries are not accessible as they are mainly under the ocean's sea level, except Iceland. Here, the main energy production is coming from geothermal sources, including also steam [21]. Eastern Africa is another example of extensional tectonics; with some, but limited use of geothermal resources. At convergent plate boundaries, especially at subduction zones, the occurrence of volcanoes manifests active geothermal systems at depth, which can be and are utilized for geothermal energy production, like in the Philippines [21]. Both resource types are usually classified as high-temperature or high-enthalpy resources, with exit temperature values of more than 150°C, which can be exploited using conventional turbine technologies where electrical power can be directly produced from hot steam or from a high-temperature two-phase fluid (steam and hot water). After electricity production, the outlet water can still reach temperatures of 150°C or less and therefore can be used for cooling and food processing and even after that for greenhouse warming, making it a cascade system [21]. In other geographical areas, like in Central Europe, for example, enhanced geothermal systems (EGS) utilize medium enthalpy geothermal resources, where much deeper reservoirs have to be tapped to get higher temperatures. Here, weakly fractured hot rocks at depth are used as energy sources, rather than directly hot water or steam. However, water has to injected from the surface through one well, and the heated water then is produced through another second well [21].

In many other countries, however, low-temperature geothermal resources can be found, with exit temperatures below 150°C, even less than 100°C. Hot springs are often the surface manifestation of such systems; they present a unique interplay of heat at depth, water circulation, geothermal reservoirs, and open pathways to the surface [22]. Geothermal reservoir temperatures of such systems, however, often high enough to be tapped as low enthalpy resources, which can be utilized for electrical energy production through binary technology systems [20]. Usually, low enthalpy resources are utilized for drying, for example, food, wool, and others, as well as heating, for example, for housing or salt evaporation [23] and not for electricity production. However, in recent years, the binary technology has rapidly advanced, for example, [24], ground installations can be quite compact and scalable (e.g., Climeon, Sweden, climeon.com; or Eavor, Canada, www.eavor.com). For such systems, the minimum feed in temperatures and flow rates currently can be as low as 70°C and 10 L/s, respectively. However, such values have to be proven through drilling geothermal exploration wells into or closer to the geothermal reservoir, a necessary requirement to ensure continues hot water supply [21].

In Southern Thailand, numerous hot springs are known and mainly used for recreational or spa activities, thus following questions arise: 1) Can the hot springs be utilized for geothermal electrical energy production? and 2) What role would geothermal energy has in a 100% renewable energy scenario for Southern Thailand?

4. Materials and methods

In Thailand, hot springs can be found from the far northern region to the South, but not in the north-eastern part of the country. Between Ratchaburi and Chumphon

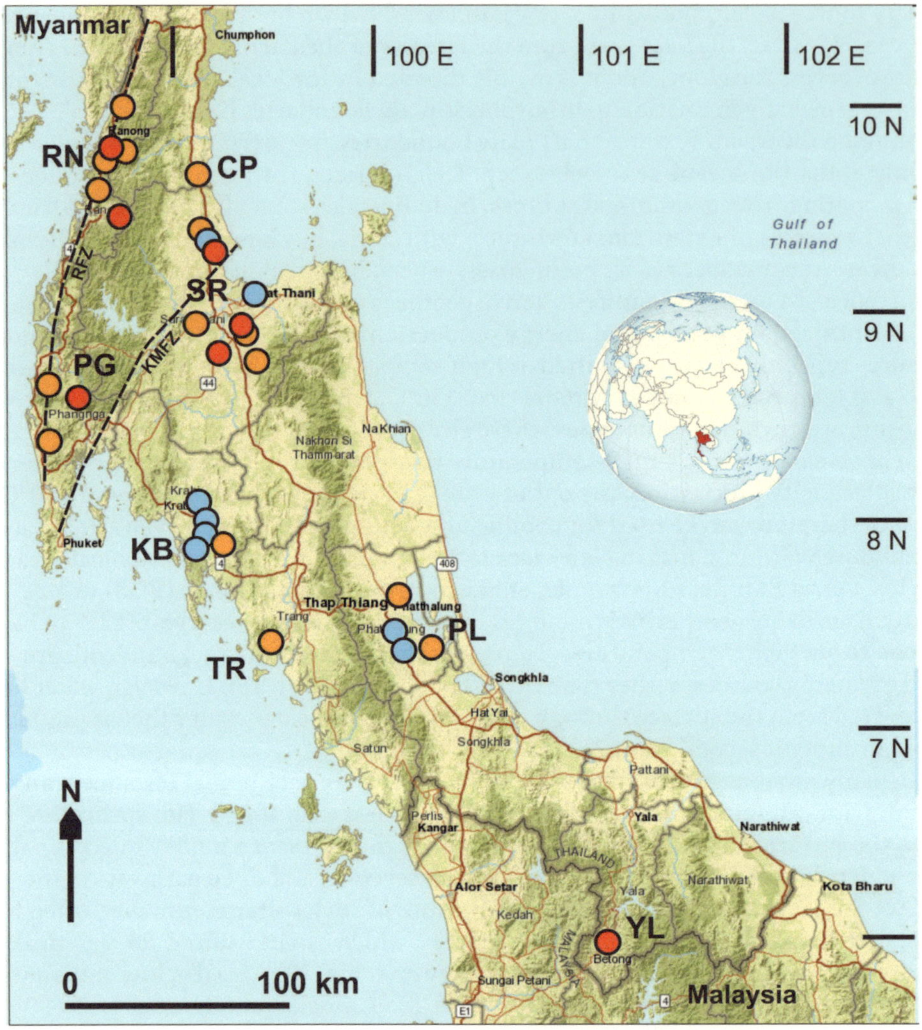

Figure 3.
Location of all main 30 hot springs in Southern Thailand (for abbreviations see text and ***Table 1****). Color code refers to* ***Figure 4****. KMFZ—Khlong Marui, and RFZ—Ranong Fault Zones. Globe via [7].*

Province, both south of Bangkok, no hot springs occur (**Figure 3**). From all hot springs in Southern Thailand, 30 main hot springs were chosen, visited, and subsequently characterized in terms of their geological setting and their water parameters.

On-site measurements have been done, and water samples have been taken at each hot spring site. Coordinates of all sampling locations were recorded using a global positioning system device (Universal Transverse Mercator, UTM, Zone 47, WGS-84). On-site measurements of exit temperatures were done in the hot spring pool and/or outflow with a glass thermometer (1°C division, max. 100°C). In all cases, water samples for geochemical analysis were collected in a 500-mL well-cleaned airtight polypropylene bottle (with good chemical resistance), after it was rinsed at least three times with the same water prior to final collection (e.g., [26]). If possible, samples were taken about 50 cm below water surface in order to prevent

Hot spring code	Location UTM (WGS-84) Zone 47		Exit temp. (°C)	Concentration (m/L)				Geo-thermo-meter (°C)
	E (m)	N (m)		Na	K	Ca	SiO_2	Quartz
CP1	512,222	1,075,014	50	63.5	6.8	89.5	64.2*	114
RN1	462,169	1,100,516	65	48.4	2.8	44.1	79.3	125
RN2	460,000	1,094,700	40	46.4	3.2	44.1	75.5*	122
RN3	461,030	1,093,400	45	46.9	3.0	44.3	72*	120
RN4	462,290	1,094,275	50	46.1	2.0	17.8	87*	130
RN5	456,192	1,080,300	46	51.3	3.5	28.1	111*	130
RN6	470,810	1,060,430	75	63.5	6.8	89.5	64.2	143
SR1	521,107	1,034,893	45	3850	132	933	65*	114
SR2	520,518	1,033,905	40	1855	64.2	400	39*	91
SR3	522,412	1,031,459	60	3655	115	840	58.5	109
SR4	555,129	1,009,502	41	20.5	2.4	27.8	37.4	89
SR5	545,897	972,938	42	55.9	6	20.9	80.2*	125
SR6	503,522	979,890	53	44.8	5.2	97.1	53.6*	105
SR7	529,417	991,895	70	64.5	13.6	381	60.7	111
SR8	530,806	991,094	56	59.7	12.8	69.6	66.3*	115
SR9	524,947	977,116	62	12.1	4.6	265	62*	112
PG1	441,455	960,807	78	84	3.3	6.9	77.7	124
PG2	437,870	975,306	55	108.6	3.1	2.6	62.5	123
PG3	420,496	918,037	45	1250	50.3	515	77	123
KB1	499,622	900,439	45	24.9	2.5	86.3	26.8	75
KB2	500,183	891,731	47	4450	125	975	44.4	96
KB3	510,462	888,220	45	986	22.1	382	25.5	73
KB4	512,329	873,475	47	12,500	395.5	833	32.1	82
KB5	523,171	876,867	47	2.3	1.3	86.3	61	111
TR1	551,391	818,787	52	79.1	2.8	82.1	61	111
PL1	625,096	823,266	57	39.1	3.7	45.5	25.5*	73
PL2	608,944	810,077	46	72.9	3.2	3	84.3*	128
PL3	604,490	816,432	50	69.7	3.2	5.1	73.6*	121
PL4	615,661	850,513	41	21.7	2.9	21.9	38.9*	90
YL1	729,730	646,758	80	76.6	6.4	16.7	97.6	136

**Data from reference [25].*

Table 1.
Code, after [25], location (in Universal Transverse Mercator, UTM, WGS-84, Zone 47), exit temperature (°C), concentrations of selected cations and anions (mg/L), and calculated reservoir temperatures (°C) of main hot spring sites in Southern Thailand.

atmospheric exposure. Bottles, labeled accordingly, indicating date, number, purpose of analysis, and location, were cooled down naturally to ambient temperatures (30–35°C) as thermal contraction is reasonably small (around 0.5% and smaller, see [27]) and then sent to the laboratory at the Faculty of Science Laboratory of Prince of Songkla University in Hat Yai. Parameters have been analyzed in a few days after samples were taken. Ca^{2+}, Na^{+}, K^{+}, and SiO_2 concentrations were determined by ICP-OES. Some data were taken from [25]. Based on the SiO_2 concentrations of the hot spring waters, silica geothermometer calculations (here quartz geothermometer) were carried out to determine the reservoir temperature [28]. All data and results are presented in **Table 1**.

5. Results

In Southern Thailand, geothermal systems can be divided into three general groups ([29]; **Figure 3**; **Table 1**): Group 1 hot springs with exit temperatures of more than 60°C are mainly found in a granitic setting, however not often in sedimentary rocks, with examples in Phang Nga (PG, Ranong (RN), and Yala (YL). The near-surface sediment layer here is quite thin. Cooler meteoric water is flowing down along open pathways, where it is heated up, and then, the hot water moves along open fractures up to the surface. Group 2 comprises hot springs with exit temperatures of around 60°C, in general associated with sedimentary or metamorphic rocks; examples here are Surat Thani (SR), Krabi (KB), Chumphon (CP), and Phatthalung (PL). The sediment cover here is comparably thicker. Also here, pathways for the cooler and hotter water are provided by open faults and fractures. However, these faults are often not fully developed up to the near surface, so that uprising hot water mixes with groundwater; this results in lower hot spring exit temperatures. Further, often more than one hot spring at the surface can be found in such areas; examples are in Surat Thani. Group 3 hot springs are associated with or are close to major fault zones. In the South of Thailand, the Khlong Marui (KMFZ) and Ranong Fault Zones (RFZ) are the main fault zones, crossing the Peninsula from the Andaman Sea to the Gulf of Thailand. Hot springs here are often directly affected by the fluid flow along such fault zones. Finally, the real heat sources for all hot springs in Southern Thailand are not yet established, either igneous bodies or higher heat flow through onshore basin development.

Surface exit temperature for the 30 hot springs in Southern Thailand ranges from 40 to 80°C, whereas the reservoir temperatures based on silica geothermometer calculations range from 73 to 143°C (**Figure 4**, **Table 1**). For seven hot springs with higher surface temperatures than 60°C (grouped in red color), the reservoir temperatures are above 105°C. For another 15 hot springs, the exit temperatures are lower than 60°C, but their reservoir temperatures are above 105°C (orange colored group). Exit temperatures lower than 60°C and reservoir temperatures lower than 105°C can be found for eight hot springs (blue-colored group). Higher exit temperatures than 60°C and reservoir temperatures lower than 105°C are not realized here.

Although no geothermal well data are available for all hot spring, the potential geothermal electrical power production can be estimated by using properties of hydrothermal fluids, for example, [30, 31], as the amount of electrical power output depends on them and the technology and type of power plant used, for example,

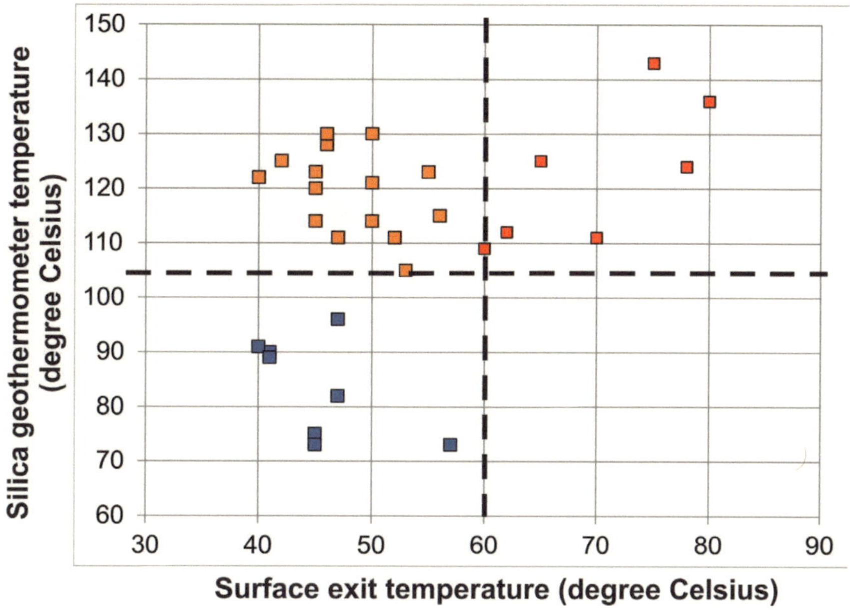

Figure 4.
Reservoir temperature based on silica geothermometer versus exit temperature of 30 hot springs in Southern Thailand. Colors relate to ***Figure 3*** *(see also* ***Table 1****).*

[30, 31]; here, binary systems utilizing a secondary fluid cycle could be suitable. Calculating generated electricity was presented with net generated electric power (NEP) [32]. Assuming power plants were running at full capacity [29], proposed Eq. (1) for calculating the approximate NEP for a binary power plant was used, considering inlet temperatures between 80°C and 150°C and outlet temperatures of 40°C and 50°C (**Figure 5**). The relative efficiency will be roughly 58% of the triangular efficiency, when adequate accuracy is required [32], with:

$$NEP \cong 2.47\dot{m}\left(\frac{T_{in}-T_0}{T_{in}+T_0}\right)(T_{in}-T_{out}) \tag{1}$$

where NEP is the approximated net generated electric power, kWe, $\dot{m}$ is the total mass, kg/s, T_0 is the dead-state temperature (20°C), T_{in} is the geothermal inlet temperature of the primary fluid, in °C, and T_{out} is the fluid temperature leaving the cold side of the heat exchanger, in °C.

The power production is highly dependent on the total mass flow rate; see Eq. (1) and **Figure 5**. Moreover, the inlet temperature of the geothermal resource also has a great effect on plant performance. At possible NEP for a site from the red and for some from the orange group is about 4 ± 0.5 MW, depending on a total mass flow rate of 25 ± 5 kg/s and a geothermal inlet temperature of about 130 ± 5°C (see also [33]). Thus, the calculation of generated power with a total mass flow of 25 kg/s lies within the expected range. A net power output of 4 ± 0.5 MW can be compared with the first geothermal plant at Apas Kiri project in Malaysia, [34, 35]. The project is set to have an installed capacity of 30 MW and will be feeding its electricity into the grid

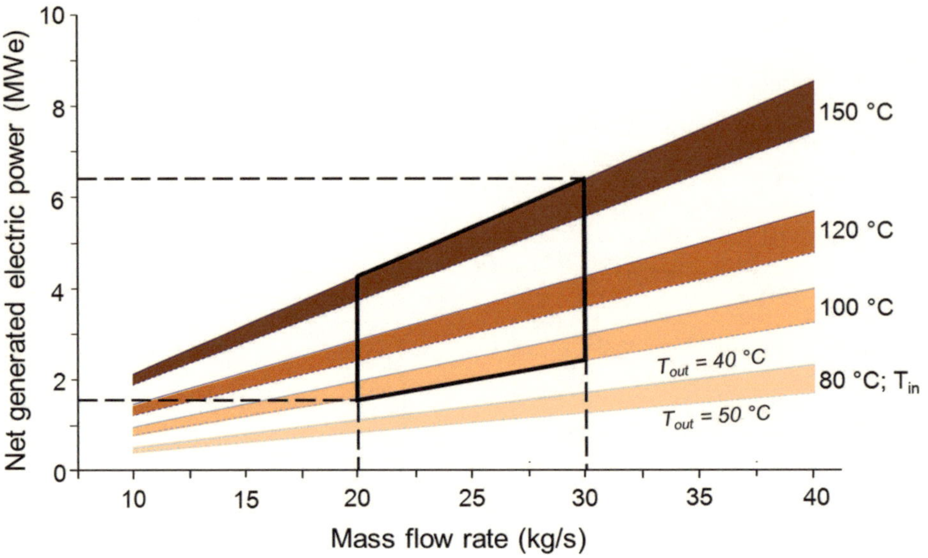

Figure 5.
Approximated net generated electrical power output for a geothermal power plant using geothermal fluids with different total mass flow rate estimates, inlet (T_{in}) and outlet temperatures (T_{out}). Dead-state temperature 20°C. for further details see text (after [33]).

of Sabah Electricity, a private limited company (www.sesb.com.my). Relative to this power plant, a geothermal power plant in Southern Thailand would be rather small and considered electricity for local scale development [36]. However, this estimate is based on theoretical and pilot plant results, and it still has to be proven commercially by drilling exploratory geothermal wells, which will provide data for mass flow rate and the water inlet temperature.

6. Discussion

6.1 State of geothermal power in Southern Thailand

The revised power development plan of 2018 aims to decrease for 2037 the share for coal-fired power from previously 23% down to 12% [37]. Natural gas was and still will be the main source of electricity generation, up from 30–53% share of overall power generation. Renewable energy sources including hydro power will increase to 29%. Nuclear power dropped out of the revised PDP as Thailand before aimed for a nuclear power plant. Some electrical energy will be imported from neighboring countries, mainly Myanmar and Laos. Geothermal energy is not listed in the PDP 2018 [34]. In the 2017 Renewable Energy Outlook for Thailand IRENA [38], it was written that "*the development of geothermal has since [1989] then been stagnant due to very little resource availability*" and further ... "*Thailand has quite modest geothermal resources with a temperature range of 40-60 °C, with some spots reaching about 80 °C. Even though the current installation of 300 kW [Fang geothermal power plant in Northern Thailand] can be upgraded or expanded to the magnitude of MW in the future, it would nonetheless remain insignificant.*" The temperatures presented in the outlook are incorrect and therefore misleading, as first, they indicate exit temperatures and not reservoir temperatures, and second, the exit temperatures, for example, in Northern Thailand, are much higher than stated in [38]. For the Fang geothermal power plant cited there, for example, fluid inlet temperatures are 110–115°C, with some wells reaching 130°C. The associated hot springs reach exit temperatures of 90–99°C, [39], in other areas even beyond.

Results of this study show that in Southern Thailand, there is a potential for geothermal electrical energy production, although for each site it has to be confirmed through well data. The main areas as shown in **Figure 3** are in Surat Thani (ST), Ranong (RN), and Phang Nga (PG), as well as Yala (YL) in the far south. However, other factors for geothermal power plants also have to be considered, for example, infrastructure, national parks, and others (see [40]). Small-scale geothermal power plants with approximately 4–5 MW to potentially 10 MW net power output would be able to provide electrical energy without any dependence on wind or solar radiation and thus would be able to provide base load to the electrical grid. Although the overall contribution would be comparable small geothermal power plants can provide stable electrical energy in rural areas where most of the hot springs are located, and it would be also in line with the Ministry of Energy's project "*Energy for All,*" [41]. According to Energy Policy and Planning Office (EPPO) deputy secretary-general Wattanapong Kurovat, "*the 2018 PDP's main objective is to ensure that each region has enough power and stable sources. It was thus important for every region to have its own base-load power plants as reliable sources,*" he said [17].

6.2 Renewable energy scenario for Southern Thailand

Currently, EGAT International, a subsidiary of the Electricity Generating Authority of Thailand, and fully owned by it, is building in Vietnam a coal-fired power plant with 1320 MW [42]. For Thailand, around 30% of the total energy comes from renewable sources means that almost 70% is still coming from conventional sources, mainly gas, but also lignite and coal, which will contribute to a continuous increase in CO_2 gas emissions, thus to still rising Earth temperatures. The 1.5°C limited defined in the Paris Agreement is far from achievable, as other countries also continue on Thailand's path. Although man-made climate change is scientifically proven, already in the last 20 years, seemingly many countries, including Thailand, believe that there is still enough time to act and also believe that their (30%) renewable energy share is sufficient [16]. The effects and consequences of a continuous CO_2 emission are clearly outlined in detail in recent IPCC reports, [4], as well as from other organization, for example, the World Meteorological Organization [43]. In Southern Thailand, the temperatures will rise to some degree that for certain time periods, it will be too hot outside (heat waves); similar conditions can recently be observed in Australia, for example, [44].

Hundred percent of renewable energy share for electrical power is possible according to [45], also for a country like Thailand. Mainly solar energy and to a lower extent, wind energy can provide the majority of the energy demand if policy frameworks are provided; both are already cheaper than conventional coal power plants [46]. As the availability of both sources is subject to changes over time energy storage systems are required. Here, dams play a key role, as well as batteries.

Recent analysis by Bloomberg NEF (BNEF) has shown that the levelized costs of electricity (LCOE) for battery storage (utility-scale lithium-ion battery storage system with four-hour duration running at a daily cycle and including charging costs) have fallen to $153/MWh in the second quarter of 2022, much lower than in 2018 with around $270, but with 8.4% slightly up compared with first quarter of 2021, due to volatilities in commodity prices [47]. According to a study from 2019 [48], solar or wind projects with added batteries capacity could already compete with coal- and gas-fired power plants over "dispatchable power," which is power whenever it is needed it can be delivered. However, despite the current increase in costs for renewable energy sources, which can be seen as a temporality, the differences in costs compared with fossil fuel power generation continue to get wider because fuel and carbon prices rise even faster [47]. According to [47], for the first half of 2022, the cost for new-built utility photovoltaic systems (with fixed axis) is lower than that for new-built coal and gas-fired power plants.

Natural gas-fired power plants are also significant contributors to CO_2 emissions, and a number of investment banks even stopped or will stop in the near future the financial support even for gas-based energy infrastructures [49]. Geothermal power plants, however, can produce electricity around the clock. The low-enthalpy system here might not be able to replace large coal or gas-fired power plants but small-scale CO_2 emission-free units can be installed at many locations, even close to a national park as shown with Fang geothermal plant in Northern Thailand [39], and with additional solar power their efficiency could be significantly increased. These plants can provide electrical power locally, and they can be connected *via* a decentralized or distributed grid, so that they can act as one of the backbones of a renewable energy system, for example, [50, 51]. The levelized cost of electricity for geothermal energy was in 2021 in the same range of coal or gas combined cycle [51], but might today be more completive due to the increase in fossil fuel prices (see above).

Further, geothermal resources can be used beyond electricity production, as direct heat use for industrial processes and also for cold generation used in storage facilities for seafood and other cold products. All such applications are based on well-established technologies and all of them having the advantage of no CO_2 emissions as well as no environmental pollution [21], when compared with coal or gas-fired power plants.

Finally, the Earth is a closed system where almost no matter comes in or goes out. In such systems, there is no waste as all systems are cyclic. Since the beginning of carbon-rich energy sources CO_2 was a waste product but not treated as one. A significant number of research studies show that CO_2 emission at all levels and in all sectors have to be taxed in order to decarbonize them [52].

7. Conclusion

Southern Thailand has geothermal resources that can and need to be tapped as part of a renewable energy mix in order to achieve zero CO_2 emissions in a foreseeable future, where every ton of CO_2 not emitted in to the atmosphere counts, and therefore to keep the global warming as low as possible. Although the geothermal resources are not of high temperature like in other countries, medium enthalpy binary technological systems are available, and further innovations and increased productions will make them also cheaper in the coming future, a trend that has been seen already in other areas, for example: in solar PV, wind turbines, and batteries.

The production of energy from renewable sources will be decentralized as outlined above; geothermal is here a good example. However, these decentralized sources then will be connected *via* data and electricity lines to customers and other renewable energy sources in order to ensure energy availability for everyone as well as spatial and temporal energy security. Solar PV will be the main source of renewable energy for Thailand, with wind being the second one. This energy scenario is possible for Southern Thailand, and also for the whole country, technologically and also economically [50]. Changes in the Power Development Plan PDP 2018 Revision 1 compared with the previous one have shown that the Thai government adheres its commitment to reduce the CO_2 emission, but only as far as it is not significantly affecting the current electricity production system, which would require a major transformation of EGAT and related companies. The latest Power Development Plan with its Revision 1 shows quite clear that among the current government and energy leadership there is not enough political will to go a path with larger CO_2 emission reduction as such a decarbonized energy system would require much more decentralization. For example, in 2026, a 600 MW lignite-fired power plant is due to replace older ones in Lampang Province, Northern Thailand [53]. Even current geopolitical (Ukraine war in 2022) and geo-economic (COVID-19 pandemic-related disruptions) factors leading to higher gas prices and recent cost advantages of renewable energy sources (see above, [44]), have not really impacted the general political course. For a global perspective, staying on the current policies means a likely increase of the global average temperatures (land and ocean) of roughly 3.0°C, according to [54], by the end of the century, and not 1.5°C, with all the consequences, especially climate tipping points [55].

Acknowledgements

The author highly acknowledges the support from W. Ngansom and also all other members of the Geothermal and Groundwater Group at Geophysics@PSU. Part of this work has been presented at the International Conference on Sustainable Energy and Green Technology 2019 11-14 December 2019, Bangkok, Thailand [56].

Conflict of interest

The authors declare no conflict of interest.

Author details

Helmut Duerrast
Faculty of Science, Geophysics Research Center, Prince of Songkla University, Hat Yai, Thailand

*Address all correspondence to: helmut.j@psu.ac.th

References

[1] Scripps Institution of Oceanography. Latest CO2 Reading. 2022. Available from: https://scripps.ucsd.edu/programs/keelingcurve/. [Accessed: July 1, 2022]

[2] NASA GISS. GISS Surface Temperature Analysis (GISTEMP v4). 2020. Available from: https://data.giss.nasa.gov/gistemp/. [Accessed: July 1, 2022]

[3] Bonfils CJW, Santer BD, Fyfe JC, Marvel K, Phillips TJ, Zimmerman SRH. Human influence on joint changes in temperature, rainfall and continental aridity. Nature Climate Change. 2020;**10**: 726-731. DOI: 10.1038/s41558-020-0821-1

[4] IPCC. The Intergovernmental Panel on Climate Change. 2022. Available from: https://www.ipcc.ch/. [Accessed: July 1, 2022]

[5] Global Carbon Project. Carbon budget and trends 2020. 2022. Available from: https://www.globalcarbonproject.org/carbonbudget. [Accessed: July 1, 2022]

[6] The Energy Policy and Planning Office. Thailand 2015 Energy Statistics of Thailand 2015. EPPO, Ministry of Energy; 2020. Available from: http://www.eppo.go.th/info/cd-2015/EnergyStatistics of Thailand 2015.pdf. [Accessed: July 1, 2022]

[7] TUBS, Globe with CC BY-SA 3.0 permission by TUBS, commons.wikimedia.org/wiki/File:Thailand_on_the_globe_(Asia_centered).svg

[8] Ramsar. Ramsar Sites Information Service: Krabi Estuary. 2020. Available from: https://rsis.ramsar.org/ris/1100. [Accessed: July 1, 2022]

[9] Bangkok Post. Protesters Rejoice after Coal 'victory'. 2018. Available from: https://www.bangkokpost.com/thailand/general/1415079/protesters-rejoice-after-coal-victory. [Accessed: July 1, 2022]

[10] Electricity Generating Authority of Thailand. Power Plants and Dams. 2020. Available from: http://www.egat.co.th/en/index.php?option=com_content&view=article&id=92&Itemid=117. [Accessed: July 1, 2022]

[11] Clean Development Mechanism. 2012, Project 2620: Srijaroen Palm Oil Wastewater Treatment Project in Krabi Province Thailand. 2020. Available from: http://cdm.unfccc.int/Projects/DB/JQA1244008061.03/view. [Accessed: July 1, 2022]

[12] Khaokoon International. GULF Commences the Operation of 25MW "Gulf Chana Green Biomass Project". 2020. Available from: https://www.kaohoon.com/content/344793. [Accessed: July 1, 2022]

[13] Energy Absolute. Wind Power Plant Production. 2020. Available from: https://www.energyabsolute.co.th/windpower.asp. [Accessed: July 1, 2022]

[14] Electricity Generating Authority of Thailand. 300 MW Thailand-Malaysia HDVC Interconnection System 2020. Available from: https://www2.egat.co.th/hvdc/INTRODUCTION.HTML. [Accessed: July 1, 2022]

[15] Electricity Generating Authority of Thailand. EGAT's Power Projects. 2016. Available from: https://www.egat.co.th/en/index.php?option=com_content&view=article&id=317&Itemid=137 [Accessed: July 1, 2022]

[16] Energy Policy and Planning Office (EPPO), Ministry of Energy, Thailand, Power Development Plan 2018. Revision 1 (PDP2018 Revision 1). 2018. Available

from: http://www.eppo.go.th/images/Infromation_service/public_relations/PDP2018/PDP2018Rev1.pdf. [Accessed: July 1, 2022]

[17] The Nation. Power Plan 'A Setback for Sustainable Energy'. 2018. Available from: https://www.nationthailand.com/national/30360098. [Accessed: July 1, 2022]

[18] The Nation. Cabinet Nods to Plan for Bt18.7 bn Economic Zone in Songkhla. 2020. Available from: https://www.nationthailand.com/business/30380887. [Accessed: July 1, 2022]

[19] IEA. Thailand 2017 Electricity Generation by Fuel. International Energy Agency (IEA). 2017. Available from: https://www.iea.org/stats/WebGraphs/THAILAND2.pdf. [Accessed: July 1, 2022]

[20] EEC. Eastern Economic Corridor. 2020. Available from: https://eng.eeco.or.th/en. [Accessed: July 1, 2022]

[21] Stober I, Bucher K. Geothermal Energy. 2nd ed. Berlin Heidelberg: Springer; 2021. p. 390. DOI: 10.1007/978-3-030-71685-1

[22] Praserdvigai S. Geothermal development in Thailand. Geothermics. 1986;**15**(5/6):565-582. DOI: 10.1016/0375-6505(86)90066-0

[23] Kaczmarczyk M, Tomaszewska B, Operacz A. Sustainable utilization of low enthalpy geothermal resources to electricity generation through a Cascade system. Energies. 2020;**13**:2495. DOI: 10.3390/en13102495

[24] Frick S, Kranz S, Kupfermann G, Huenges E. Making use of geothermal brine in Indonesia: Binary demonstration power plant Lahendong/Pagolombian. Geotherm Energy. 2019;**7**:30. DOI: 10.1186/s40517-019-0147-2

[25] Department of Mineral Resources. Hot Springs in Thailand. Report. 2012. Available from: https://www.dmr.go.th/. [Accessed: July 1, 2022]

[26] Huang W-J, Wang Y, Cai W-J. Assessment of sample storage techniques for total alkalinity and dissolved inorganic carbon in seawater. Limnology and Oceanography: Methods. 2012;**10**:711-717. DOI: 10.4319/lom.2012.10.711

[27] Arnórsson JO, Bjarnason N, Giroud N, Gunnarsson I, Steffánsson A. Sampling and analysis of geothermal fluids. Geofluids. 2006;**6**:203-216. DOI: 10.1111/j.1468-8123.2006.00147.x

[28] Fournier RO, Rowe JJ. Estimation of underground temperatures from the silica content of water from hot springs and wet- steam wells. American Journal of Science. 1966;**264**:685-697. DOI: 10.2475/ajs.264.9.685

[29] Raksaskulwong M. Thailand geothermal energy: Development history and current status. In: Proceedings of the 8th Asian Geothermal Symposium; 9-12 December 2008; Hanoi. Vietnam. 2008. pp. 39-46

[30] Zarrouk SJ, Moon H. Efficiency of geothermal power plants: A worldwide review. Geothermics. 2014;**51**:142-153. DOI: 10.1016/j.geothermics.2013.11.001

[31] Bertani R. Geothermal power generation in the world 2005-2010 update report. Geothermics. 2012;**41**:1-29. DOI: 10.1016/j.geothermics.2011.10.001

[32] DiPippo R. Ideal thermal efficiency for geothermal binary plants. Geothermics. 2007;**36**:276-285. DOI: 10.1016/j.geothermics.2007.03.002

[33] Ngansom W. Geothermal Resources in Southern Thailand: Integrated Geoscientific Investigations and

Assessments [Thesis]. Hat Yai: Prince of Songkla University; 2018

[34] Barnett PR, Mandagi S, Iskander T, Abidin Z, Armaladdoss A, Raad R. Exploration and development of the Tawau geothermal project, Malaysia. In: Proceedings of the World Geothermal Congress; 19-25 April 2015; Melbourne, Australia.

[35] Chong LHH, Mohd ND. Tawau Hill Park springs, Sabah, Malaysia. GHC Quart Bull. 2000;**21**(4):3-4

[36] Rubio-Maya C, Ambríz Díaz VM, Pastor Martínez E, Belman-Flores JM. Cascade utilization of low and medium enthalpy geothermal resources – A review. Renewable and Sustainable Energy Reviews. 2015;**52**:689-716. DOI: 10.1016/j.rser.2015.07.162

[37] The Diplomat. Having a renewable energy transition is a critical step to realize Thailand 4.0. 2019. Available from: https://thediplomat.com/2019/03/thailands-renewable-energy-transitions-a-pathway-to-realize-thailand-4-0/. [Accessed: July 1, 2022]

[38] IRENA. Renewable Energy Outlook: Thailand. International Renewable Energy Agency (IRENA). 2017. Available from: https://www.irena.org/publications/2017/Nov/Renewable-Energy-Outlook-Thailand. [Accessed: July 1, 2022]

[39] Wood SH, Kaewsomwang P, Singharajwarapan FS. Geologic framework of the fang Hot Springs area with emphasis on structure, hydrology, and geothermal development, Chiang Mai Province, northern Thailand. Geothermal Energy. 2018;**6**:3. DOI: 10.1186/s40517-017-0087-7

[40] Ngansom W, Duerrast H. Assessment and ranking of Hot Springs sites representing geothermal resources in southern Thailand using positive attitude factors. Chiang Mai Journal of Science 2019;46(3):592-608. DOI: http://cmuir.cmu.ac.th/jspui/handle/6653943832/66035

[41] Bangkok Post. Energy for All Scheme to Open at End of the Month. 2020. Available from: https://www.bangkokpost.com/business/1931012/energy-for-all-scheme-to-open-at-end-of-the-month. [Accessed: July 1, 2022]

[42] Bangkok Post. EGAT International Builds Coal-Fired Power Plant in Vietnam. 2019. Available from: https://www.bangkokpost.com/business/1799979/egat-international-builds-coal-fired-power-plant-in-vietnam. [Accessed: July 1, 2022]

[43] WMO. State of the Global Climate in 2021. World Meteorological Organization; 2022. Available from: https://public.wmo.int/en/our-mandate/climate/wmo-statement-state-of-global-climate. [Accessed: July 1, 2022]

[44] ABC News. Heatwave update: Temperatures are expected to peak over the coming days. 2019. Available from: https://www.abc.net.au/news/2019-01-16/summer-heatwave-expected-across-australia/10719356. [Accessed: July 1, 2022]

[45] Ram M, Bogdanov D, Aghahosseini A, Gulagi A, Oyewo AS, Child M, Caldera U, Sadovskaia K, Farfan J, Barbosa LSNS, Fasihi M, Khalili S, Dalheimer B, Gruber G, Traber T, De Caluwe F, Fell H-J, Breyer C. Global Energy System Based on 100% Renewables. Power, Heat, Transport and Desalination Sectors Study by LUT University & Energy Watch Group (Lappeenranta, Berlin). 2019. Available from: http://energywatchgroup.org/wp-content/uploads/-EWG_LUT_100RE_All_Sectors_Global_Report_2019.pdf. [Accessed: July 1, 2022]

[46] The Two-Way. Peabody Energy, a Giant in the Coal Industry, Files for Bankruptcy. 2016. Available from: https://www.npr.org/sections/thetwo-way/2016/04/13/474059310/u-s-coal-giant-peabody-energy-files-for-bankruptcy. [Accessed: July 1, 2022]

[47] BloombergNEF. Cost of New Renewables Temporarily Rises as Inflation Starts to Bite. 2022. Available from: https://about.bnef.com/blog/cost-of-new-renewables-temporarily-rises-as-inflation-starts-to-bite/. [Accessed: July 1, 2022]

[48] BloombergNEF. Battery Power's Latest Plunge in Costs Threatens Coal. Gas. 2019. Available from: https://about.bnef.com/blog/battery-powers-latest-plunge-costs-threatens-coal-gas/. [Accessed: July 1, 2022]

[49] European Investment Bank. EU Bank Launches Ambitious New Climate Strategy and Energy Lending Policy. 2019. Available from: https://www.eib.org/en/press/all/2019-313-eu-bank-launches-ambitious-new-climate-strategy-and-energy-lending-policy.htm. [Accessed: July 1, 2022]

[50] Jacobsen MZ. The cost of grid stability with 100% clean, renewable energy for all purposes when countries are isolated versus interconnected. Renewable Energy. 2021;**179**:1065-1075. DOI: 10.1016/j.renene.2021.07.115

[51] Lazard. Lazard's Levelized Cost of Energy Analysis – Version 15.0. 2021. Available from: https://www.lazard.com/media/451881/lazards-levelized-cost-of-energy-version-150-vf.pdf. [Accessed: July 1, 2022]

[52] Carbon Tax Center. Recommended Policy Journals and Papers. 2019. Available from: https://www.carbontax.org/contact-us/recommended-policy-journals-and-papers/. [Accessed: July 1, 2022]

[53] EGAT: 23 September 2021 NEB greenlights Mae Moh Power Plant Replacement Project and Surat Thani Power Plant Project. 2021. Available from: https://www.egat.co.th/en/news-announcement/news-release/neb-greenlights-mae-moh-power-plant-replacement-project-and-surat-thani-power-plant-project. [Accessed: July 1, 2022]

[54] Hausfather Z, Peters GP. Emissions – The 'business as usual' story is misleading. Nature. 2020;**577**:618-620

[55] McKay DIA, Staal A, Abrams JF, Winkelmann R, Sakschewski B, Loriani S, et al. Exceeding 1.5°C global warming could trigger multiple climate tipping points. Science. 2022;**377**:6611. DOI: 10.1126/science.abn7950

[56] Duerrast H. Geothermal resources in southern Thailand – Part of a renewable energy mix. In: IOP Conference Series: Earth and Environmental Science, International Conference on Sustainable Energy and Green Technology 2019; 11-14 December 2019. Bangkok, Thailand: IOP 463; 2020. p. 012146

Chapter 3

Toward Sustainable Implementation of Geothermal Energy Projects – The Case of Olkaria IV Project in Kenya

Lilian Namuma S. Kong'ani and Raphael M. Kweyu

Abstract

In this chapter, we demonstrate how geothermal has the potential to solve climate change. Geothermal is part of green energy, which contributes toward the achievement of sustainable development goals, that is, SGD 7, on affordable, reliable, sustainable, and modern energy for all, SDG 13, on climate actions, and the Paris Agreement. We present the potential of geothermal energy in Kenya and link it to its ability to provide solutions for Africa and Kenya considering current geopolitics, including Brexit, climate change, the Russian-Ukraine war, and COVID-19. However, this chapter argues that geothermal energy production should be developed within a sustainability framework. Environmental conflicts occasioned by the implementation of developmental projects are on the rise. Geothermal projects are likely to introduce new conflicts between the government and the communities. Therefore, natural resource conflict resolution should be part of the development of geothermal energy. This chapter draws inspiration from a study on conflict types and their management in the Olkaria IV geothermal development project in Kenya. From the study, it is apparent that mediation is one of the sustainable environmental conflict management strategies. The chapter concludes that geothermal energy production has the potential to contribute to the prosperity of Kenya economically.

Keywords: conflict management, geothermal energy development, involuntary resettlement, project affected persons, sustainability

1. Introduction

Geothermal energy is increasingly being taunted as one of the essential resources in fighting the worrisome climate change worldwide. Increased calls for a need to expedite addressing climate change continue to dominate the headlines in different forums globally, including in the United Nations Climate Change Conference (COP26) held in 2021 in Glasgow, United Kingdom. During the COP26 conference, 34 countries and 5 public finance institutions pledged to redirect their public support from fossil fuels to renewable energy.

Notably, concerted efforts have been made worldwide toward investment in the exploitation of renewable energy, including geothermal, to reduce carbon footprints. The development of the geothermal industry enables the availability of one of the most reliable renewable energies that are naturally extracted from the earth's crust. Globally, installed geothermal energy production hit 15,608-megawatt electric (MW) by 2021 [1], with the top three countries, including the United States contributing over 3714 MW, Indonesia about 2233 MW, and the Philippines contributing 1918 MW [1]. Regionally, the Great East African Rift is among the most important world regions harboring a significant geothermal potential of more than 15,000 MW [2], with about 67% of this potential being in Kenya [1, 3].

Kenya tops the African nations in terms of geothermal power generation and is one of the fastest-growing geothermal power producers in the world. The installation of the Olkaria V geothermal power plant (172 MW) in November 2019 pushed the country's geothermal production capacity up to 865 MW with more than 35% of the households in Kenya depending on geothermal power [4]. Currently, Kenya has overtaken Iceland (755 MW), to rank eighth worldwide [1, 5]. The country is nearing the ranks of the United States, Indonesia, Philippines, Turkey, and New Zealand, which are in club 1GW following the commissioning of Olkaria I unit 6 with an installed capacity of 83 MW, which pushes the total geothermal power generation to 944 MW as at 2022.

Geothermal energy development is playing a fundamental role in the energy market in Kenya, contributing about 50% of total generated electricity in 2020/2021 [3, 4]. This is followed by hydro at 39% and thermal at 15%, a drop from 32% in the first half of 2021, while a mere 0.4% is derived from wind power [1].

The exploration of geothermal energy in Kenya seeks to enable the transition of the country into a newly industrialized, middle-income state by 2030, and provide a high quality of life to all its citizens in a clean and secure environment [6]. Geothermal exploitation validates Kenya's global commitment toward inter alia, the Sustainable Development Goals (SDG), that is, SDG 7 on affordable, reliable, sustainable, modern energy for all, and SGD 13 on climate actions [6] as well as the Paris Agreement. SDGs 7 and 13 are vital to the realization of other SDGs. These include SDG 1, on ending poverty in all forms, SDG 2, on eliminating hunger as well as SDG 3, on improving health and wellbeing [7] among others.

The energy sector in Kenya is one of the crucial forces behind its economy, which is key in the manufacturing and agriculture sectors. These sectors are a key backbone to the country's economic growth. Yet, the energy industry is hit hard by a myriad of global and local challenges, including climate change, global pandemics, and social and political instabilities, such as the Russian-Ukraine war and community opposition among others, resulting in energy scarcity with increased prices. Higher oil prices, for instance, escalates production costs, which are subsequently borne by consumers, further resulting in increased cost of living and continued reliance on biomass energy and fossil fuels.

However, a global energy crisis exacerbated by global issues is a blessing in disguise. They present an opportunity for countries, such as Kenya, to pursue and intensify investments in the locally available renewable energy, such as geothermal, wind, solar, tidal, wave, and hydro energy, which remain largely untapped, with only about 9% of geothermal energy exploited from its potential of 10,000 MW [3, 4], as at 2022. Locally, adequate public participation in the design and implementation of energy projects is important in navigating community opposition menace for the sustainability of the projects. As a result, Kenya would be better placed to accelerate the harnessing of renewable energy resources and cut down on its reliance on energy importation and related costs. These resources would be injected into more impactful public costs and address other socio-economic challenges in the country.

2. Development of geothermal energy

Geothermal can be explained as the heat from the earth's crust estimated to be about 5500°C at the core of the earth, which is as hot as the sun's surface [8]. This energy is manifested on the surface of the earth in form of hot springs, fumaroles and hot-altered sites. Geothermal is the only renewable energy source created naturally by the earth. Geothermal energy is harnessed from underground reservoirs, consisting of hot water and steam, which are naturally replenished, making it both renewable and sustainable.

Deep wells, that is about two kilometers, are drilled to access hot water and steam from the underground reservoirs, and piped up to a well, where it is used to drive turbines connected to electric generators. This creates power for various uses in industries and homes, such as lighting and heating up buildings. In Kenya, geothermal energy is also used for direct utilization at the geothermal fields, including Olkaria, Menengai, and Eburru. These uses include fish farming, recreational purposes, pasteurization of milk, drying of crop harvests, and heating and fumigation of greenhouses [1, 9, 10].

Geothermal energy is a clean, renewable resource that can be tapped globally by countries, such as Kenya, which are located in geologically favorable areas. Geothermal energy is deemed a renewable resource due to the exploitation of the heat from the interior of the earth, which is considered abundant. The used hot water and steam can be cooled and channeled back to the reservoir.

2.1 Advantages of geothermal energy

Geothermal energy advantages over other sources of power, such as wind, solar, and hydro, include:

2.1.1 Stability

Geothermal energy is not affected by the disruption caused by unfavorable weather conditions, such as droughts. It has the highest availability, which is estimated at over 90%, especially in Kenya [11, 12]. Thus, more reliable and secure, and a more suitable source for baseload electricity generation in the country [13, 14].

2.1.2 Eco-friendly

Geothermal is green energy with minimal adverse effects on the environment. Geothermal fields have a low carbon footprint since the energy is extracted from the earth without burning fossil fuels. The pollution associated with geothermal energy is relatively minimal compared to other fossil fuels, such as coal, natural gas, and crude oil.

2.1.3 Vast potential

Increased investment and research toward the exploitation of geothermal resources, with accelerated new technologies, enabling the use of untapped reservoirs. This has contributed to the accessibility, efficiency, and application of geothermal energy to a wider range of uses. Currently, the advancement in the geothermal energy extracting process, with new technologies, enabling the extraction of geothermal energy from deeper reservoirs.

2.1.4 Small land footprint

Geothermal energy is extracted from the earth's crust, thus can be established on small pieces of land compared to solar, wind, and hydropower energy, which requires large parcels of land. The national geographic estimates that about 400 square miles of the land surface would be adequate to establish a geothermal power plant capable of producing 1 GW-hour of electricity, while a solar and wind farm at the same energy output would need about 2340 and 1335 square miles, respectively.

2.2 Disadvantages of geothermal energy

2.2.1 Restricted location

The installation of geothermal energy plants is restricted to specific locations. Most large geothermal plants require geothermal reservoirs above 100°C, which can only be found near tectonic plate boundaries or hot spots [15], such as the East African Rift System, characterized by the presence of quaternary volcanic centers, that are younger than approximately 2.6 million years, along the rift's margin, with younger centers situated in the south and older centers farther north.

2.2.2 Greenhouse gas emissions

The extraction of geothermal energy from the earth's surface leads to the release of greenhouse gases, such as hydrogen sulfide, carbon dioxide, methane, and ammonia. While these gases are also released into the atmosphere naturally, the rate increases near geothermal plants. However, emissions of these gases are significantly lower than those associated with fossil fuels.

2.2.3 Earthquakes risk

Geothermal power plant installations involve drilling deep within the earth to release hot steam and/or water trapped in rock formations. This causes alterations in the structure of the earth and instability underground that can lead to earthquakes on the earth's surface. Geothermal wells collapse has been reported in the 1950s and 1960s in Wairekei, New Zealand [16]. Geothermal power plants have the potential to cause slow land subsidence over time as geothermal reservoirs are depleted. However, the implications of earthquakes are minor since most of the geothermal plants are situated away from communities.

2.2.4 High costs

Exploration of geothermal energy is capital intensive. A 50 MW well drilling could cost about USD 180 million at testing through full scale development [1]. However, upon its full implementation, the well could be operational for up to 40 years, enabling the recouping of the initial costs.

2.2.5 Summary of the merits and demerits of geothermal energy

Table 1 presents an overview of the advantages and disadvantages of geothermal energy.

2.3 Geothermal resources development in Kenya: a history in brief

Geothermal resources in Kenya are found within the rift valley, which forms part of the East Africa Rift System (EARS), with an estimated potential of up to 10,000 MW spread over 14 potential sites. The EARS is connected to the worldwide oceanic rift systems of over 30 million years ago. The rifting events resulted in tectonic shifts and volcanism and geothermal activity are associated with the occurrence of quaternary volcanoes located within the rift's axis.

Kenya's geothermal exploration for power generation began in 1952 led by the then East African Power & Lighting Company Ltd (EAPL), with the support of the United Nations Development Program (UNDP) and other international agencies [17, 18]. The study resulted in the drilling of two wells in the 1950s. Although temperatures of up to 235°C were recorded, the wells were only discharged in 1971 after stimulation.

Later, the Olkaria geothermal area was selected by the studies that were commissioned to evaluate the resources in various sectors of the rift for a thorough evaluation. This led to the drilling of six deeper exploration and appraisal wells in Olkaria, which were successfully completed and proved the existence of a viable geothermal system. Thus, the first geothermal power plant, Olkaria I, with an electric power capacity of 45 MW, was constructed between 1981 and 1985 (**Table 2**).

Currently, the Olkaria geothermal fields, which are second-most productive in the world after the geysers field in the USA, host five power plants [19], including Olkaria I-V commissioned in the years 1981, 2003, 2009, 2014, and 2019, respectively, with plans to construct Olkaria VI and VII [19–21]. The installed geothermal capacity comprises 706.8 MW by Kenya Electricity Generating Company (KenGen), 155 MW by OrPower4, Inc and 3.6 MW by Oserian Development Company Ltd. Further, 45 MW was added to the grid by Orpower4 between 2015 and 2018. 45 Inc. Olkaria geothermal field is currently the main producing site with an installed capacity of 689.7 MW, while Eburru field has an installed capacity of 2.52 MW.

Merits	Demerits
Geothermal energy is stable/reliable. It is not dependent on prevailing weather conditions.	Energy is confined to specific locations where hot water and/or steam can be tapped from the earth's crust.
It emits minimal greenhouse gases (GHGs) unlike other fossil fuels, thus environmental friendly.	The drilling of the resource triggers the release of GHGs into the atmosphere contributing to global warming.
Abundant renewable energy. New technologies, research, and investment accelerate its exploitation from untapped reservoirs.	Drilling of the geothermal resource causes instability underground with the potential to cause earthquakes.
Energy is extracted from within the earth's surface, thus a small land footprint is required vis-à-vis other renewable energy sources, such as solar and wind.	High upfront cost in the exploration of this energy.

Table 1.
Merits and demerits of geothermal energy.

Station and licensee	Year commissioned	Installed capacity	Status
Olkaria I, KenGen	Unit 1 (1981)	15 MW	Generation and production drilling
	Unit 2 (1982)	15 MW	
	Unit 3 (1985)	15 MW	
	Unit 4 (2014)	70 MW	
	Unit 5 (2015)	70 MW	
	Unit 6 (2022)	83 MW	
		Total = 185 MW	
Olkaria II, KenGen	Unit 1 (2003)	35 MW	Generation and production drilling
	Unit 2 (2003)	35 MW	
	Unit 3 (2010)	35 MW	
Olkaria III, Orpower4	Unit 1 (2000)	48 MW (total)	Generation and production drilling
	Unit 2 (2009)	36 MW	
	Unit 3 (2014)	26 MW	
	Unit 4 (2016)	29 MW	
		Total = 139 MW	
Olkaria IV, KenGen	2014	140 MW	Generation and production drilling
Olkaria V, KenGen	2019	2 × 82.7 MW = 165.4 MW	Generation and production drilling
Olkaria VI, KenGen	2022 (expected)	140 MW	Surface exploration and production drilling
Suswa, CYRQ Energy	2024 (expected)	2 × 37.5 MW	Surface exploration and production drilling
		6 × 42.5 MW	
		Total = 330 MW	
Eburru, KenGen	Unit 1 (2012)	2.5 MW	Generation and Pilot generation
	Unit 2 (2019, expected)	22.5 MW	
		Total 25 MW	
Akira, AGL[a]	2022 (expected)	1 × 70 MW	Exploration and surface studies
Oserian, ODCL[b]	2003	2.5 MW	Production under steam sale
Longonot, AGIL[c]	2019 (expected)	140 MW	Production drilling
Bogoria-Silali, GDC[d]	2021 (expected)	200 MW	Production drilling
Menengai, GDC	2020 (expected)	3 × 35 MW	Production and exploration drilling
		Total = 105 MW	

[a]*Akiira Geothermal Limited.*
[b]*Oserian Development Company Limited.*
[c]*African Geothermal International Limited.*
[d]*Geothermal Development Company.*
Source: Energy & Petroleum Regulation Authority, Kenya.

Table 2.
Geothermal energy fields and status of development in Kenya.

The commissioning of the public-private partnership (PPP), 140 MW power plant, the Olkaria 1 unit 6, 83.3 MW, and 105 MW power plant, which is under development at Menengai geothermal field are expected to increase geothermal power development in Kenya by 328 MW between 2020 and 2022. The Menengai project intends to involve the Geothermal Development Company (GDC) as a steam supplier, while three independent power producers (IPPs) will each install 35 MW. While KenGen continues to appraise and develop several sectors of the Olkaria field, the GDC has further mobilized a drilling rig for exploration drilling in the Paka prospect, and also intends to drill exploratory wells in Silali, Korosi, and the Greater Menengai field within the next few years.

Thirteen IPPs have since been licensed by the government of Kenya to undertake greenfield (areas that have not been previously been developed) projects at Barrier, Longonot, Akiira, Elementaita, Homa Hills, Menengai North, Lake Magadi, Arus, Baringo, Emuruangogolak, Namarunu, and Emuruapoli prospects. These efforts demonstrate Kenya's quest to increase the country's geothermal output to 5000 MW from the current 944 MW by 2030 [1].

2.4 Barriers to geothermal development in Kenya

The East African Rift System has significant potential for clean energy exploitation for the countries in East Africa, Kenya included. Yet, over 95% of the geothermal energy resources remain unexploited. Similarly, to other African states, Kenya is facing a number of challenges in maximizing the harnessing of geothermal resources. These issues included.

2.4.1 High exploitation and infrastructure costs

Whereas, steam or hot water is readily available for constant supplies at Olkaria geothermal field, the likely delays in exploitation experienced elsewhere in the rift valley by private companies in Longonot and Akiira demonstrate the difficulty in finding investors who would be patient to finance additional exploration. A single exploration well where no previous development has been done costs over USD 1 million to drill, with three wells needed to prove resource availability [22]. It is also estimated that a 20 MW geothermal power plant could cost about USD 80 million, which could be unaffordable in the event of a reduced number of customers with declined demand.

2.4.2 Political instability and community opposition

Political instability and community opposition are major deterrence to development and investment in geothermal resources, especially for IPPs. The geothermal development in Olkaria IV, for instance, faced community resistance following its relocation in 2014 amidst claims of unfair compensations, which almost derailed its implementation. However, the application of mediation as a conflict resolution strategy, in this case, helped to reduce conflicts between the developer and the project affected persons (PAPs), mended relationships, improved community livelihoods, and allowed smoother operations of the project.

2.4.3 National and county levels bureaucracy

The control at the national levels is often deemed as a threat to the county governments, which fail to adequately manage their own issues including ensuring that "*Wanjiku,*" that is, the local communities at the county levels are well represented [22] in all matters of development including energy projects. Also, added bureaucracies at the national levels are seen as fertile ground for political interferences and a threat to approvals for important projects, such as geothermal energy. The bureaucracies at the county levels, with possible inadequate participation of the private developers in the management of local affairs, are considered as a possible avenue for corruption, which could adversely impact important development.

2.5 Opportunities for the development of the geothermal industry in Kenya

2.5.1 Prevailing global issues

The escalation in energy prices emanates from the weakening of economies already battered by the impacts of the coronavirus pandemic worldwide, such as lockdowns and disturbances, to global supply chains worsened by increased fuel prices. The Russian-Ukraine war has heightened the energy crisis, further resulting in uncertainty in global oil and gas markets with soaring energy prices.

In Kenya, this impact has been felt by its citizens who have had to dig deep into their pockets to meet the cost of basic necessities. For example, before the onset of the Russian-Ukraine, a 6 kg cooking gas cylinder retailed at about USD 7. This shot up to about USD 13 during the Russian-Ukraine war era. However, these global disturbances present an opportunity for countries to accelerate the transition to alternative sources of energy, including geothermal energy. This is particularly in countries, such as Kenya, which has the potential of up to 10,000 MW, yet only about 9.4% has been tapped as of 2022.

Further, the ravaging impacts of droughts with declined hydropower generation, compounded with a decline in fossil fuels, such as coal, oil, and natural gas, provides an opportunity for countries, such as Kenya, to intensify investments toward the exploitation of the untapped geothermal reservoirs.

2.5.2 Resource availability

Kenya is endowed with abundant geothermal resources, which are estimated at about 10,000 MW. These resources are found along the world-famous East African Rift Valley, which transects from the north to the south of the country. The resources are spread over 14 sites, with Olkaria, Menengai, and Eburru being the most developed geothermal sites. Suswa, Longonot, Arus-Bogoria, Lake Baringo, Korosi, Paka, Lake Magadi, Badlands, Silali, Emuruangogolak, Namarunu, and Barrier are other potential sites currently under exploration. Only 944 MW has been harnessed as of 2022, enabling the country to rank eighth globally in terms of geothermal energy production.

2.5.3 Legal and institutional framework

The Energy Act, 2019, which repealed the Energy Act, 2006, the Kenya Nuclear Electricity Board Order No. 131 of 2012, and the Geothermal Resources Act, 1982 promotes renewable energy and promotes exploration, recovery, and commercial utilization of geothermal energy among others, creating an enabling environment for accelerated development of geothermal resources in the country.

Over two decades ago, the development of geothermal resources was solely tasked to Kenya Power and Lighting Company (KPLC), a state-owned electricity generation and distribution company, under the Ministry of Energy, which derailed its development. Subsequent reforms within the energy sector attracted wider energy developers. For example, the policy on new feed-in-tariffs (FIT) that was introduced in 2008 in line with the Energy Act, 2006, then provided for investment security to renewable electricity generators, reduce administrative and transaction costs attracting many IPPs, who currently include ORMAT, Akiira Geothermal Company Ltd, and Quantum East Africa Power Ltd among others into geothermal development for electricity and direct utilization in the country [17].

The Kenya Vision 2030 launched in 2008 emphasizes the exploitation of renewable energy to reduce reliance on imported fossil fuels and increase access to electricity. With this economic blueprint, the country aims to achieve a geothermal production capacity of up to 5000 MW by 2030.

2.5.4 Technical expertise

Unlike other east African countries, Kenya, currently ranked 8th worldwide in terms of geothermal energy production, has a robust, skilled, local geothermal workforce, and technical capacity. This is boosted by the establishment of the African Geothermal Training Center (AGCE) in the country. This facility was established in June 2018 by the African Union Commission and UN Environment (UNEP). However, there is still a need for support from foreign experts to maximize harnessing of the geothermal resources.

2.6 Analytical/theoretical framework

2.6.1 Sustainable practices in energy development

Sustainability Development involves a progressive transformation of economy and society. A development path that is sustainable in a physical sense could theoretically be pursued even in a rigid social and political setting. But physical sustainability cannot be secured unless development policies pay attention to such considerations as changes in access to resources and in the distribution of costs and benefits. Even the narrow notion of physical sustainability implies a concern for social equity between generations, a concern that must logically be extended to equity within each generation (Our common future, UN).

The term sustainability is simply the capacity to endure when applied broadly it can be defined as "meeting the needs of the present generation without compromising the ability of future generations to meet their own needs." Sustainable development can be defined as "Development that meets the needs of the present without compromising the ability of future generations to meet their own needs." Sustainability may be viewed as a three-legged table consisting of the environment, the economy and society, or as a dualistic relationship between human beings and the ecosystem they inhabit.

2.6.2 Environmental sustainability

This is a condition of balance, resilience, and interconnectedness that allows human society to satisfy its needs, while neither exceeding the capacity of its supporting ecosystems to continue to regenerate the services necessary to meet those needs nor by our actions diminishing biological diversity. Renewable energy production systems are largely seen to fulfill the environmental sustainability condition.

2.6.3 Economic sustainability

It involves creating economic value out of whatever project or decision you are undertaking. Economic sustainability means that decisions are made in the most equitable and fiscally sound way possible while considering the other aspects of sustainability. From an economic standpoint, sustainability requires that current economic activity not disproportionately burdens future generations. Economists will allocate environmental assets as only part of the value of natural and manmade capital, and their preservation becomes a function of overall financial analysis.

Economic sustainability should involve analysis to minimize the social costs of meeting standards for protecting environmental assets but not for determining what those standards should be. Components of the economic environment include residents and households, public infrastructure, community facilities and the natural environment (essential services, such as water and sanitation systems, electricity, gas, telecommunications, and transport), business enterprises and supply networks (retailers, distributors, transporters, storage facilities and suppliers that participate in the production and delivery of a particular product), not-for-profit sector, and government. Comparing geothermal energy to fossil fuels, the former is seen to be more economically sustainable considering the ongoing global challenges, such as the Russian-Ukraine crisis and instability in Gulf states.

2.6.4 Social sustainability

This is based on the concept that a decision or project promotes the betterment of society. Further, future generations should have the same or greater quality of life benefits as the current generation do. This concept also encompasses aspects of human rights, environmental law, and public involvement and participation. Energy production systems whether renewable or nonrenewable are socially sustainable if they fulfill the following aspects [23]:

a. Equity of access to key services.

b. Equity between generations.

c. A system of relations valuing disparate cultures.

d. Political participation of citizens, particularly at a local level.

e. A sense of community ownership.

f. A system for transmitting awareness of social sustainability.

g. Mechanisms for a community to fulfill its own needs where possible.

h. Political advocacy to meet needs that cannot be met by community action.

Environmental conflicts are generally viewed as an outcome of production systems that fail to fulfill the condition of social sustainability. The Global Environmental Justice Atlas (EJAtlas) [24], linked slightly more than 2520 socio-environmental conflicts to large projects and communities worldwide. More than 345 of these conflicts are related to the construction of renewable energy amenities, climate fixes, and dams. In Kenya, in the northern-western, Turkana County, the oil and wind projects generated conflicts between the host communities and operating companies. The communities were displeased with the unfulfilled pledges concerning land compensation, improved water supply, and employment prospects [25–28]. These concerns were exacerbated by communication break down between Tullow and the residents [26, 29]. This led to continued disruption of the company's operations [27]. Also, the fencing of all sites, including for extraction and oil storage, was fenced, restricted communities' access, and dislocated pastoral migration routes, resulting in conflicts between the developers and these communities.

The community was discontent with the benefit-sharing arrangement and accused the national government of lack of transparency in awarding the tender to Fenxi company in 2011 derailing its implementation, in the case of the Mui Basin coal exploration project located in Kitui, Kenya [30, 31]. The community was still contesting the project by the time [31] were conducting their research on public participation in Africa's mining sector.

The proposed 1050 MW Lamu coal power plant project, which was expected to be operational in 2020, had been projected to be the largest in east Africa and the first in Kenya. However, the project failed to start off in October 2015 as planned [32, 33] following the Civil Society Organization Save Lamu's and the community's opposition. These groups were anxious over unavoidable environmental and health impacts, such as pollution of fishing grounds, that would have seen hundreds of fishermen lose jobs and premature deaths linked to air pollution. The continued protests compelled the project donor, the Industrial and Commercial Bank of China (ICBC), to withdraw its financial support due to looming environmental and social hazards.

3. Conflicts and development of renewable energy in Kenya: case study of Olkaria IV geothermal energy project

This section is based on a study by [21, 34] on conflict types and management in the development of geothermal energy in Olkaria IV area. The Olkaria IV geothermal project is located in the Olkaria geothermal block in Naivasha-Sub-County, Nakuru County, Kenya, partially within the Hell's Gate National Park (HGNP). Olkaria area is inhabited by about 20,000 pastoralists, whose main livelihood stream is supported by pastoralism and livestock trading, with a number of community members relying on tourism activities (**Figure 1**) [21, 36].

Olkaria IV geothermal power plant has an installed capacity of 140 MW. The plant was established by KenGen. It was supported financially by the Government of Kenya (22%), the European Investment Bank (EIB, 12%), the Japan International Cooperation Agency (JICA, 23%), the French Development Agency (AFD, 15%), the German Development Agency (KfW, 7%), the World Bank (7%) with KenGen injecting 14% [36, 37].

Olkaria IV geothermal power plant was established as part of the Kenya Electricity Expansion Project (KEEP) to deliver on Vision 2030 of seeing Kenya transition into a newly industrialized, middle-income state, and provide a high quality of life to all

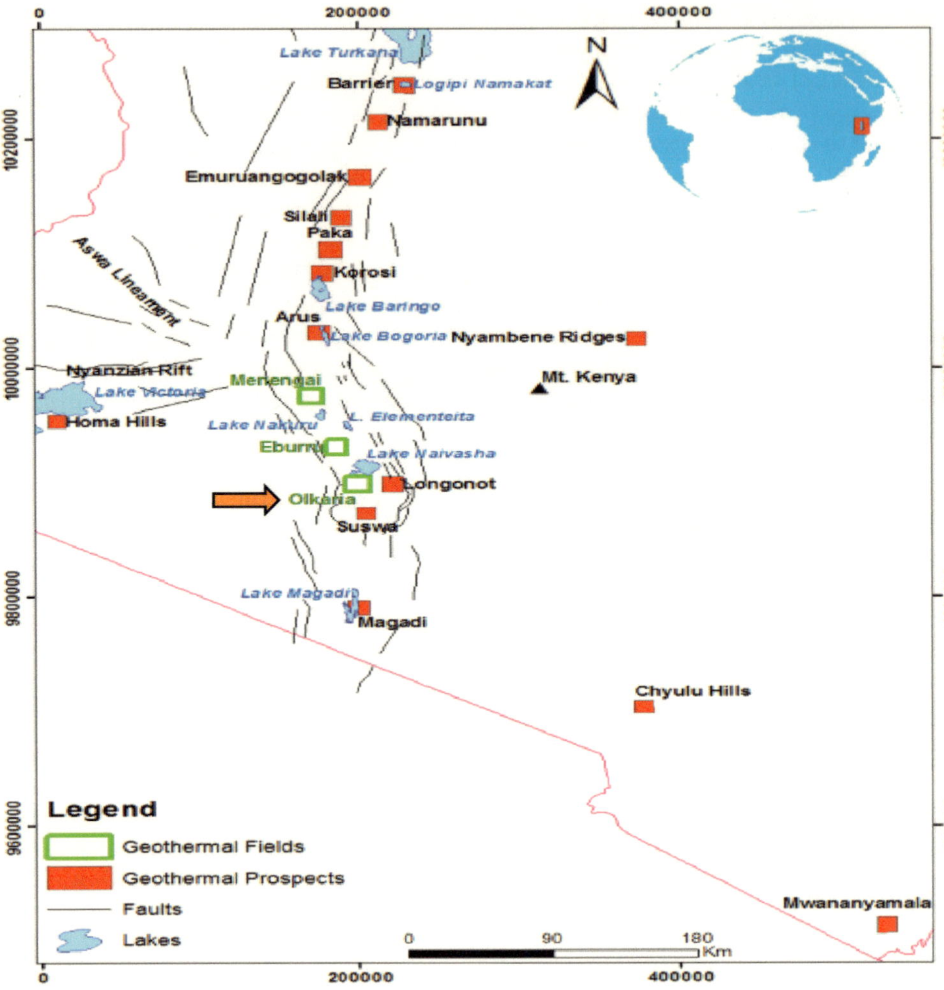

Figure 1.
Location of Olkaria geothermal field/study area. Source: [35].

its citizens in a clean and secure environment, and SDG 7 on affordable, reliable, sustainable, modern energy for all, and SGD 13 on climate actions [6]. However, its installation was faced with conflicts between KenGen and the project-affected persons (PAPs) that persisted beyond its completion.

An environmental social impact assessment (ESIA) on the project demonstrated that the drilling of the power plant would negatively impact the health of the community. This necessitated the relocation of four villages inhibited by the Maasai community, including Cultural Centre, OlooNongot, OlooSinyat, and OlooMayana Ndogo to a new site, that is, resettlement action plan land (RAPland), which was located outside the park.

However, upon relocation, the community became agitated and raised complaints regarding incomplete projects at RAPland and accused KenGen of failing to deliver on some of the pledges made earlier, as stipulated in the memorandum of understanding signed between KenGen and the PAPs. These conflicts revolved around the socio-economic (51%), cultural (14%), environmental (21%), and political (14%) aspects [21].

Regarding the socio-economic conflict, the PAPs cited increased distance to work at the project site and shopping centers in Kamere and Naivasha, and increased travel costs exacerbated by bad roads and inadequate means of transport. They also pointed out that the water collection and watering points were inadequate and unreliable, with an unreliable electricity supply, while some houses had no electricity. Declined income accrued from selling traditional ornaments and guided tours at the former site was another main cause of disagreement. The respondents suggested that they would have appreciated adequate financial compensation, including USD 5000, as a disturbance allowance to help them settle in the new site.

Environmentally, the respondents complained of poor terrain characterized by poor grazing areas of low-quality pasture. The unwelcoming valleys and gullies posed a danger to community members and their livestock, while the hyenas had become a nuisance, killing PAPs' livestock daily. Also, the respondents were unconvinced of the development's potential adverse effects on their health as earlier informed by KenGen. They claimed that they were never furnished with documented scientific evidence of the latent negative effects of noise pollution, as indicated by the developer.

Cultural issues were mainly linked to the standard two-bedroom houses built at RAPland, which some PAPs claimed failed to provide for the customary needs for exclusive units for the husbands, wives, sons, and daughters. Also, some women were dissatisfied on the basis of their views being disregarded, exacerbated by the patriarchal system, which forbids women from speaking in the same public spaces as men. The community leaders had an obligation to make decisions on behalf of the community. Thus, community members had to abide by a decision made regarding relocation irrespective of their feelings.

Politically, KenGen was accused of improper sharing of information relating to the development of the project. Also, the developer was blamed for the alleged inadequate involvement of PAPs in project meetings and in the decision-making processes involving their compensation and relocation logistics. PAPs felt that KenGen tricked them to relocate, so as to expand geothermal developments by making promises, some of which were never fulfilled, including a USD 5000 disturbance allowance. The majority of the PAPs (77%) would have appreciated more support, including financial compensation and more time to prepare for the relocation.

3.1 Geothermal energy production conflict effects

Conflicts resulted in abandoned businesses at HGNP and reduced tourism activities. This impacted negatively on the livelihoods of members of the Cultural Center

Approach	Indicator	Project phase
Competition	PAPs were coerced into agreement	Initiation, relocation
Avoidance	PAPs involuntarily agreed to relocate	Relocation
Collaboration	Attempts were made for KenGen and PAPs to talk and resolve issues through meetings/*barazas*	Initiation, relocation, implementation
Accommodation	PAPs had no choice but to move to pave the way for the establishment of Olkaria IV for the benefit of the entire nation.	Initiation, relocation
Compromise	Mediation and negotiation	Implementation

Table 3.
Conflict management approaches.

village, who led in these activities. It was also noted that about half of PAPs lost their jobs through punitive measures taken for participating in protests against relocation, while those who resisted relocation lost friends. PAPs that were seen to associate themselves with the resistance group were threatened with legal sanctions and isolation by the developer.

3.2 Geothermal energy production management of conflicts

Traditionally, conflict can be explained as "a struggle over values and claims to scarce status, power and resources in which the aims of the opponents are to neutralize, injure or eliminate their rivals [38, 39]." Generally, conflicts exist wherever or whenever incompatible activities occur, including in developmental projects, and may result in win-lose character. Flagship projects, such as geothermal energy, which bring together diverse stakeholders, including the host community, the state, the project developers, and donors, among others, often attract conflicts following their varying interests. Conflict occurrences can also be heightened by the factors, such as different comprehension of the project plans, resource scarcity, and varying priorities of the stakeholders involved [40, 41].

Unresolved conflicts can have detrimental implications in a development project, including hurting the relationships, between the developer and the community with subsequent delays in project implementation, loss of the project's social license, increased cost of the project, its rejection, termination, and in worst-case scenario, loss of lives [21, 42–45].

This study documented the various strategies used to manage the conflicts associated with geothermal energy production in Olkaria, Naivasha Sub-county, Nakuru County, and Kenya. The main strategies employed were competition, avoidance, collaboration, compromise, and accommodation (**Table 3**). The different strategies were employed at different stages of the project implementation.

However, competition, avoidance, collaboration, and accommodation as conflict management strategies were deemed ineffective following the persistence of conflicts beyond relocation. The PAPs wrote to the World Bank and the European Investment Bank seeking their intervention, leading to mediation that was recommended by the project donors. This mediation process was fruitful, as indicated by 82% of the interviewees. The PAPs applauded the mediated negotiation of the twenty-seven thorny issues inter alia, the construction of five more houses for those who had been left out, and improved services at RAPland, most of which were agreeably addressed. This has since then led to an improved relationship between KenGen and the community.

4. Conclusion

This chapter has attempted to position geothermal energy production in Kenya as a sustainable development practice. Indeed, Kenya has great potential for green energy based on its strategic geographical location. Kenya's physical environment provides a great opportunity for green energy, that is, solar and wind energy potential in vast northeastern arid lands, wind, tidal, and waves energy at the coast, and geothermal energy springs in the rift valley. Kenya heavily relies on fossil fuels for its production and transport systems. Most of the imported energy for Kenya comes from the Gulf states. However, with current global geopolitics, such as Brexit, the Russian-Ukraine war, and conflicts in areas, such as Yemen, the supply and prices for fossil fuels have become very unstable and unpredictable. Consequently, there is an alarming increase in prices of commodities that heavily rely on fossil fuels in their production and transportation lines. This has resulted in economic inflation and the situation has been exacerbated by the recent COVID-19 pandemic, which has slowed economic activities globally. The current situation is threatening to plunge Kenya into an economic crisis with threats of political instability occasioned by unrest by the masses who are not only unable to get employment opportunities but also to put food on their table thanks to high food prices. It should also be noted that the world is moving toward green energy as a way of combating climate change caused by the increase in greenhouse gases emitted into the atmosphere, hence leading to global warming. European countries are already setting targets for going green. Germany, for example, targets to have all cars on its roads electric by 2030. This global developmental shift to green energy is important for African countries, and Kenya in particular. Kenya is a leading geothermal energy producer in Africa and has great unharnessed potential. Kenya can position itself as an exporter of green energy (geothermal, wind and solar, etc.) to the rest of Africa if the right investment decisions are made now. This chapter argues that the right energy development decisions are those that conform to the three pillars of sustainable development, namely, the economy, ecology, and society. Whereas geothermal energy production is arguably one of the most economically and environmentally friendly, there is a need to be sensitive to the needs and aspirations of the communities within which the wells are developed to achieve social sustainability. One of the most sustainable environmental conflict management strategies is mediation as has been demonstrated by ref. [21, 34] in the case of conflict resolution between the government and Maasai communities in the Olkaria IV geothermal field.

Acknowledgements

The authors wish to greatly thank Prof. R. G. Wahome and Dr. Thuita Thenya who were the co-authors for the case study used in this chapter. We also wish to thank anonymous reviewers and informants for any role played toward the case study.

Author details

Lilian Namuma S. Kong'ani[1*] and Raphael M. Kweyu[2]

1 Department of Earth and Climate Sciences, University of Nairobi, Nairobi, Kenya

2 Department of Geography, Kenyatta University, Nairobi, Kenya

*Address all correspondence to: lnamuma@uonbi.ac.ke

References

[1] ThinkGeoEnergy. Kenya Geothermal Energy Market Overview. Kenya: Think GeoEnergy; 2021

[2] Kombe EY, Muguthu J. Geothermal energy development in East Africa: Barriers and strategies. Journal of Energy Research and Reviews. 2019;**2**:1-6

[3] EPRA. Energy & Petroleum Statistics Report. Kenya: Energy & Petroleum Regulatory Authority; 2021

[4] EPRA. Energy & Petroleum Statistics Report. Kenya: Energy & Petroleum Regulatory Authority. 2020. Available from: https://www.epra.go.ke/wp-content/uploads/2021/03/Energy-and-Petroleum-Statistics-Report-2020.pdf

[5] UNESCO. UNESCO Science Report: The Race Against Time for Smarter Development. 2021. UNESCO Publishing, Paris: The United Nations Educational, Scientific and Cultural Organization. Available from: https://unesdoc.unesco.org/in/documentViewer.xhtml?v=2.1.196&id=p::usmarcdef_0000377433&file=/in/rest/annotationSVC/DownloadWatermarkedAttachment/attach_import_07223302-8f4a-4e99-9997-d370ea8d1818%3F_%3D377433eng.pdf&locale=en&multi=true&ark=/ark:/48223/pf0000377433/PDF/377433eng.pdf#%5B%7B%22num%22%3A2656%2C%22gen%22%3A0%7D%2C%7B%22name%22%3A%22XYZ%22%7D%2C-1%2C842%2C0%5D

[6] The Energy Act. Nairobi: The Government Printer; 2019

[7] Ouedraogo NS. Opportunities, barriers and issues with renewable energy development in Africa: a comprehensible review. Current Sustainable/Renewable Energy. 2019;**6**:52-60

[8] Barasa K, Moses J. Geothermal electricity generation, challenges, opportunities and recommendations. International Journal of Advanced Science and Research. 2019;**5**:53-95

[9] Mangi PM. Geothermal exploration in Kenya - status report and updates. In: SDG Short Course III on Exploration and Development of Geothermal Resources, organized by UNU-GTP and KenGen. Lake Bogoria and Lake Naivasha, Kenya. 2018. Available from: https://orkustofnun.is/gogn/unu-gtp-sc/UNU-GTP-SC-27-0701.pdf

[10] Wangari V. Direct Uses of Geothermal Resources in Menengai, Kenya. Nairobi, Kenya. 2020. Available from: https://theargeo.org/C8/final/Direct%20uses%20of%20Geothermal%20Resources%20in%20Menengai.pdf

[11] Merem EC, Twumasi Y, Wesley J, et al. Analyzing geothermal energy use in the East African region: the case of Kenya. Energy Power. 2019;**9**:12-26

[12] Simiyu SM. Status of geothermal exploration in Kenya and future plans for its development. In: World Geothermal Congress. Bali, Indonesia. 2010. pp. 25-29

[13] Kunze C, Hertel M. Contested deep geothermal energy in Germany—the emergence of an environmental protest movement. Energy Research and Social Science. 2017;**27**:174-180

[14] Pan S-Y, Gao M, Shah KJ, et al. Establishment of enhanced geothermal energy utilization plans: barriers and strategies. Renewable Energy. 2019;**132**:19-32

[15] Limberger J, Boxem T, Pluymaekers M, et al. Geothermal energy in deep aquifers:

a global assessment of the resource base for direct heat utilization. Renewable and Sustainable Energy Reviews. 2018;**82**:961-975

[16] Bolton RS, Hunt TM, King TR, et al. Dramatic incidents during drilling at Wairakei Geothermal Field, New Zealand. Geothermics. 2009;**38**:40-47

[17] Omenda P, Mangi P, Ofwona C, et al. Country Update Report for Kenya 2015-2019. 2021. Available from: https://pangea.stanford.edu/ERE/db/WGC/Abstract.php?PaperID=4876

[18] Ouma PA. Geothermal exploaration and development of the Olkaria geothermal field. 2009. Available from: https://orkustofnun.is/gogn/unu-gtp-sc/UNU-GTP-SC-10-1102.pdf

[19] Renkens I. The impact of renewable energy projects on indigenous communities in Kenya. The cases of the Lake Turkana Wind Power project and the Olkaria Geothermal Power plants. No. 28, The International Work Group for Indigenous Affairs (IWGIA)

[20] Koissaba BRO. Geothermal energy and indigenous communities: the Olkaria projects in Kenya. Heinrich Böll Stiftung European Union. 2018. file:///C:/Users/ACER/Downloads/geothermal-energy-and-indigenous-communities-olkariaproject-kenya%20(1).pdf

[21] Kong'ani LNS, Wahome RG, Thenya T. Variety and management of developmental conflicts: the case of the Olkaria IV geothermal energy project in Kenya. Conflict, Security and Development. 2021;**21**:781-804

[22] Johnson OW, Ogeya M. Risky business: Developing geothermal power in Kenya. Stockholm Environment Institute. 2018:8

[23] Morelli J. Environmental sustainability: a definition for environmental professionals. Journal of Environmental Sustainability. 2011;**1**:1-10

[24] Temper L, Demaria F, Scheidel A, et al. The global environmental justice atlas (EJAtlas): ecological distribution conflicts as forces for sustainability. Sustainability Science. 2018;**13**:573-584

[25] Schilling J, Locham R, Weinzierl T, et al. The nexus of oil, conflict, and climate change vulnerability of pastoral communities in northwest Kenya. Earth System Dynamics. 2015;**6**:703-717

[26] Schilling J, Weinzierl T, Lokwang AE, et al. For better or worse: Major developments affecting resource and conflict dynamics in northwest Kenya. Zeitschrift für Wirtschaftsgeographie. 2016;**60**:57-71

[27] Schilling J, Locham R, Scheffran J. A local to global perspective on oil and wind exploitation, resource governance and conflict in Northern Kenya. Conflict, Security and Development. 2018;**18**:571-600

[28] Vasquez PI. Kenya at a crossroads: Hopes and fears concerning the development of oil and gas reserves. Poldev. 2013;**4**

[29] Johannes EM, Zulu LC, Kalipeni E. Oil discovery in Turkana County, Kenya: a source of conflict or development? African Geographical Review. 2015;**34**:142-164

[30] Neumann M. Extractive industries and the poor in Africa - a case study of coal mining in the Mui Basin, Kenya [thesis]. Radboud University Nijmegen; 2015. Available from: https://theses.ubn.ru.nl/bitstream/handle/123456789/3959/Neumann%2C_Martin_1.pdf?sequence=1

[31] Omondi J, Karanja W, Wairimu & Co, et al. Public Participation in Africa's Mining Sector. 2020. Available from: https://www.extractiveshub.org/servefile/getFile/id/7639

[32] Banktrack. Banktrack. Lamu coal power plant Kenya. 2020. Available from: https://www.banktrack.org/project/lamu_coal_power_project

[33] Boulle M. The hazy rise of coal in Kenya: the actors, interests, and discursive contradictions shaping Kenya's electricity future. Energy Research and Social Science. 2019;**56**:101205

[34] Kong'ani LNS, Wahome RG, Thenya T. Managing geothermal project implementation conflicts through mediation: a case of Olkaria IV Project, Nakuru county, Kenya. Journal of Sustainability, Environment and Peace. 2022;**5**:96-108

[35] Munyiri SK. Structural Mapping of Olkaria Domes Geothermal Field using Geochemical Soil Gas Surveys, Remote Sensing and GIS [thesis]. United Nations University; 2016. https://orkustofnun.is/gogn/unu-gtp-report/UNU-GTP-2016-05.pdf

[36] Schade J. Kenya 'Olkaria IV' Case Study Report: Human Rights Analysis of the Resettlement Process. 2017. p. 199

[37] Abad A. Conclusions Report. Complaint SG/E/2014/07 Complaint SG/E/2014/08 Olkaria I and IV Kenya, European Investment Bank - Complaints Mechanism. Available from: https://www.eib.org/attachments/complaints/sg-e-2014-07and-08-conclusions-report-en.pdf

[38] Lawal RO, Orunbon NO, Ibikunle GA, et al. Resolving conflict in African traditional society: An imperative of indigenous African system. Euro Afro Studies International Journal. 2019;**1**:38-55

[39] Otite O, Albert IO, editors. Community conflicts in Nigeria: management, resolution and transformation. Abuja: Spectrum Books; 1999

[40] Liu JY, Low SP. Work–family conflicts experienced by project managers in the Chinese construction industry. International Journal of Project Management. 2011;**29**:117-128

[41] Wu G, Zhao X, Zuo J, et al. Effects of contractual flexibility on conflict and project success in megaprojects. International Journal of Conflict Management. 2018;**29**:253-278

[42] Batel S, Devine-Wright P, Tangeland T. Social acceptance of low carbon energy and associated infrastructures: a critical discussion. Energy Policy. 2013;**58**:1-5

[43] Arthur J, Pia L, Solveig M. Local acceptance of wind energy: factors of success identified in French and German case studies. Energy Policy. 2007;**35**:2751-2760

[44] Karytsas S, Polyzou O, Mendrinos D, et al. Towards social acceptance of geothermal energy power plants. In: European Geothermal Congress. Den Haag, The Netherlands. 2019. p. 6

[45] Wei H-H, Liu M, Skibniewski MJ, et al. Conflict and consensus in stakeholder attitudes toward sustainable transport projects in China: an empirical investigation. Habitat International. 2016;**53**:473-484

Chapter 4

Environmental and Socio-Economic Impact of Deep Geothermal Energy, an Upper Rhine Graben Perspective

Eléonore Dalmais, Guillaume Ravier, Vincent Maurer, David Fries, Albert Genter and Béatrice Pandélis

Abstract

The Upper Rhine Graben is a region renowned in Europe for the exploitation and development of geothermal energy with projects in France, Germany and Switzerland. In the last 20 years, numerous seismic events have been felt by local population triggering social concerns that have been addressed at different levels (state regulation, technical adaptation of projects and communication). Indeed, geothermal projects need a high level of acceptance by inhabitants in the surrounding area. In this regard, the local socio-economic impact is a crucial factor in social acceptance. Nevertheless, this energy resource has many advantages such as competitive heat prices and low environmental impacts, quantified by Life Cycle Analysis. This approach is also completed by continuous environmental monitoring. Moreover, additional valorization of geothermal water through its use for low temperature heating or recovery of mineral resources are ways of providing additional benefits to the local community. This chapter is dedicated to present the environmental and socio-economic impacts of two operational EGS projects (Soultz-sous-Forêts and Rittershoffen) located in Northern Alsace (France) producing geothermal electricity and heat in a rural area.

Keywords: enhanced geothermal system, induced seismicity, life cycle analysis

1. Introduction

Geothermal development in the Upper Rhine Graben (URG) involves a geothermal doublet system consisting in a production well with a down-hole pump and an injection well which reinjects cold water into the geothermal reservoir. Thus, they consist of two deviated wells that crosscut a local permeable normal fault or fracture zone in which geothermal brines are circulating by thermal convection [1] (**Figure 1**). Typical production and injection temperatures in the URG range from 150–170°C to 60–80°C.

Over the last 20 years, several deep geothermal energy projects in Europe experimented with enhancing initially low reservoir permeability based on

IntechOpen

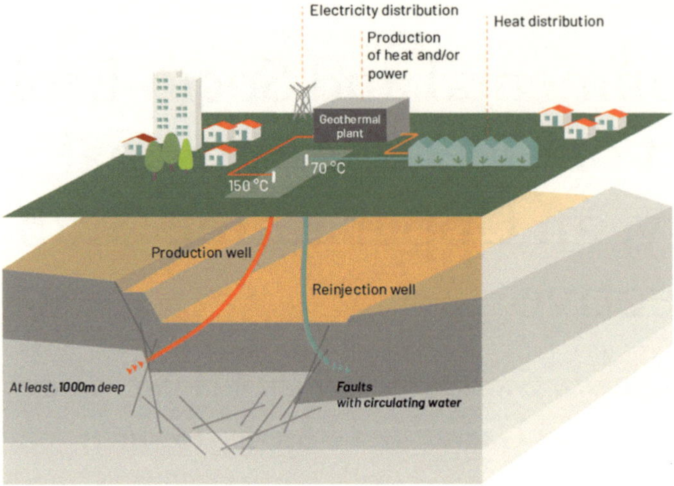

Figure 1.
Schematic of a generic deep geothermal project in the URG showing a doublet structure (production and reinjection wells) in a naturally permeable faulted reservoir.

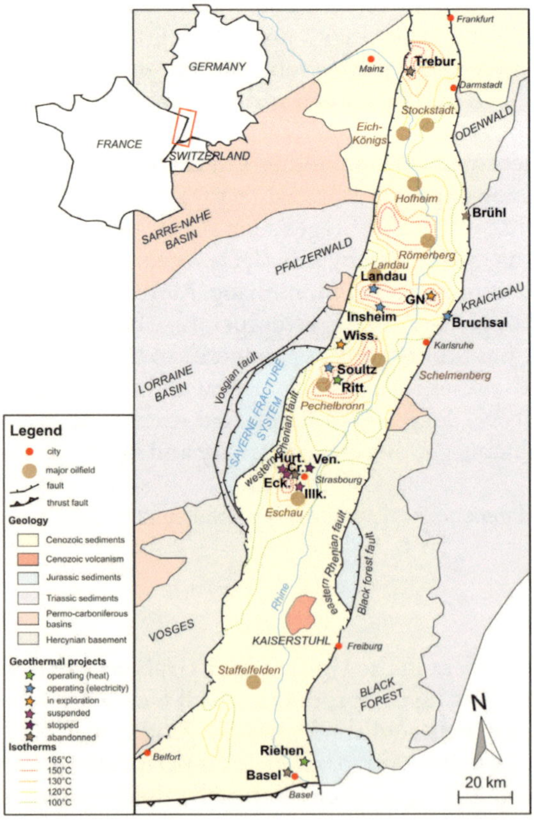

Figure 2.
Map of deep geothermal projects in the upper Rhine graben, modified after [2]; project abbreviations: Cr.: Cronenbourg; Eck.: Eckbolsheim; GN: Graben-Neudorf; hurt.: Hurtigheim; Illk.: Illkirch; Ritt.: Rittershoffen; Ven.: Vendenheim; and Wiss.: Wissembourg.

DOI: http://dx.doi.org/10.5772/intechopen.107395

various stimulation techniques. Those projects known as Enhanced or Engineered Geothermal System (EGS) are mainly located in the Upper Rhine Graben in France (Soultz-sous-Forêts, Rittershoffen, Vendenheim, Illkirch), Germany (Landau, Insheim, Bruchsal) and Switzerland (Basel, Riehen) (see **Figure 2**). The Riehen project located in the Eastern part of the URG, is not considered as a EGS project but as a hydrothermal project [1].

This chapter presents the environmental and territorial impacts of these geothermal projects. It focuses on induced seismicity and how the operators and mining authorities have introduced operational limitations to mitigate the seismic risk. Additionally, it reviews the other risks to the local environment and their mitigation practices in comparison to the environmental benefits of energy resource. Furthermore, it highlights the economic impact of the development of geothermal energy and mineral extraction on the local area. This chapter is mostly illustrated with the French experience of EGS in the URG but also includes German and Swiss examples where relevant.

2. The upper Rhine graben, a transborder region where deep geothermal energy is commonly associated with seismic risk

In the URG, natural water infiltrates and circulates through convection loops up to the interface between sediments and the basement due to a natural network of faults and fractures [3]. To obtain an economically viable flow rate in the production and injection wells, various techniques may be applied to enhance the well connection to the fractured reservoir, through thermal, chemical and/or hydraulic stimulation techniques which qualify them as EGS [4]. Induced microseismicity can occur in the vicinity of the well due to the reinjection of the water in the fractured reservoir: due to a hydromechanical mechanism [5], but also a thermal effect [6].

The examples listed below focused on deep geothermal energy and the associated induced seismicity which occurs during hydraulic stimulation phases on average after only a few days of technical operation. Geothermal sites in the exploitation phase that represent more than at least two decades of continuous operations are also presented below. The drilling phases and related potential nuisances are not considered here given that there are generally, no seismic events during drilling operations.

2.1 Impact of massive injection during hydraulic stimulation

2.1.1 Soultz-sous-Forêts site

Even though the Soultz (Soultz-sous-Forêts) wells were stimulated hydraulically and chemically several times from the 1990s onwards [7] only a few examples of felt seismicity have occurred during hydraulic stimulations at Soultz in 2000 and 2003. The most striking episode corresponds to a massive hydraulic stimulation carried out in the Soultz fractured granite reservoir in 2003 at a depth of 5 km with a wellhead overpressure of around 170 bar, when an induced seismic event was felt with a magnitude MD of 2.9 (MD—Magnitude Duration) [8]. In that specific case, experts from the site owners and the local population's insurers were mobilized, but were unable to prove any concrete structural damage to housing. Therefore, no damage was actually caused by this event, but in the minds of local residents, acceptability became a real issue. Thus, the site's operators are continuing to develop this site whilst minimizing hydraulic stimulation and explaining and communicating in-depth with local stakeholders such as politicians or associations.

2.1.2 Basel geothermal site

A deep geothermal well drilled at Basel in a fractured granite reservoir at 5 km depth in the southern part of the URG was hydraulically stimulated by massive injection with a well-head over pressure of 300 bar. This event caused structural damages and led to the permanent shutdown of this project [9]. The Basel project is considered as a counter reference for EGS development because an earthquake of magnitude ML > 3.4 (ML—Magnitude Local) was felt during a hydraulic stimulation in 2006 [10].

2.1.3 Vendenheim geothermal site

In the Strasbourg area, two deviated wells were drilled to a depth of about 5 km in a fractured granite reservoir in an urban area. A series of man-made earthquakes was felt between 2019 and 2021 in the Strasbourg area and on the German side induced by a complex sequence of hydraulic injection involving both high well-head overpressure and high cumulative massive volume [11]. The second largest event (ML 3.6), induced on 4 December 2020, led to the project being permanent shut down, but further events continued to be felt later, in 2021, with the largest one reaching a maximum magnitude of ML > 3.9 felt on the surface on 26 June 2021 [12]. As a result of the structural damage observed on many houses, the Prefect of Strasbourg decided to suspend all geothermal activity in this urban area.

2.2 Impact of geothermal exploitation on induced seismicity

2.2.1 Landau geothermal site

The Landau geothermal plant is made up of two deviated wells drilled to a depth of about 3.5 km in fractured granite. Hydraulic and chemical stimulations were successfully conducted to improve permeability without any felt seismicity [13]. However, during geothermal exploitation, an induced event was felt on 15 August 2009 with MD = 2.7. Moreover, from 2013 to 2014, a technical incident occurred in the injection well inducing an uplift of several centimeters in the geothermal platform [14]. Thus, some damage potentially caused by the uplift was observed. The injection well was repaired and the geothermal exploitation restarted some years later.

2.2.2 Rittershoffen, Soultz-sous-Forêts and Insheim sites

Some other geothermal projects, such as Rittershoffen are considered as geothermal success story because no induced earthquakes were felt during their development phase (thermal, chemical or hydraulic stimulation) [4], or during geothermal exploitation [15]. The maximum local magnitude recorded during the on-site development phase on was 1.7 [15]. This plant has now been operating for 6 years and has a capacity of more than 24 MW of heat at highest flow rate in the URG (i.e. more than 80 L/s).

At Soultz, no induced seismic event have been felt after more than 6 years of exploitation [16]. The same observations are made for the Insheim geothermal power plant in Germany [17].

2.3 Impact of felt seismic events on deep geothermal energy development

2.3.1 Differing risk perception towards new projects

Following the shutting down of the Vendenheim project in 2020, the Illkirch project located in the southern part of Strasbourg was also suspended by the French mining authorities even though no induced event related to this site had been felt. Therefore, only one deviated well was drilled in fractured granite to a depth of 3.3 km, and this geothermal site is now on standby awaiting a decision from the mining authorities for it to go ahead. The Prefect of Strasbourg mandated a group of scientific experts to provide a better appreciation of the seismic behavior of the deep reservoir and to find out why such sequence of felt events took place. Nevertheless, it is clear that the series of seismic events at Vendenheim drastically affected the perception of geothermal energy by the local population in the Strasbourg's area.

Despite these seismic events, some projects have continued their development such as the Riehen geothermal heat plant. This plant extracts geothermal water at a depth of 1500 m in fractured Triassic limestones to deliver 20 L/s of water at 65°C. It has been providing thermal energy to 8500 residents since 1994. Despite its very close proximity to Basel, the two city centers being just a few kilometers apart, this plant has not been affected by the overall distrust shown towards geothermal energy. Indeed, in Autumn 2020, in a timeframe coinciding with seismic events in Strasbourg, a plan to expand this geothermal plant was put to a local referendum and accepted by the population. To enhance social acceptance, the Riehen and Basel local utility companies, developing this project, set up transparent communication mediated by an independent third party, the Risiko-Dialog foundation (https://www.risiko-dialog.ch/en/geothermie-im-dialog/, July 2022).

On the German side of the Upper Rhine Graben, projects developed after 2015 changed the drilling target from the deep fractured granite basement to the shallower fractured Triassic sediments (mostly sandstones) overlying the basement. This change in target was accompanied by a communication drive to highlight the lower seismic risk associated with such reservoirs. Currently, in 2022, these projects are still in development, and further observations will be necessary in the upcoming years to assess this strategy. On the French side, a similar approach has been developed since 2020 with projects targeting a relatively shallower depth (down to max. 3500 m) compared to that of Vendenheim or Basel (around 5000 m). Furthermore, the mining authorities are requesting more detailed seismic risk analyses as part of the process of examining applications for authorizations to drill geothermal wells.

These examples highlight the fact that, in spite of the various counter examples cited above, the public perception of the seismic risk associated with deep geothermal energy in the URG is dependent on many factors.

2.3.2 How local population could contribute as a stakeholder to geothermal development in Alsace?

The first acceptability survey conducted in Alsace was carried out in 2012 among the local populations living close to the Soultz power plant, including the two villages of Soultz-sous-Forêts and Kutzenhausen [18]. This study, which involved the mayors of the two villages, demonstrated that the Soultz plant was well accepted by the local population even if there were some complaints about some technical nuisances such as the noise generated by the geothermal plant or reservoir management drawbacks such as induced seismicity.

At that time, geothermal energy was perceived as a favorable technology by the local population, even if there were some people who were reticent. In conclusion, the risks related to geothermal exploitation were reasonably well accepted by residents in the surrounding area.

More recently, as part of the EU DESTRESS project, social scientists have conducted an acceptability study comparing the public perception of geothermal energy in a rural area like Northern Alsace, where geothermal energy is accepted, and in an urban area like Strasbourg, where geothermal projects have raised some issues [19]. It turned out that locally "anchored" projects implemented by companies with local roots are much better perceived by the public than "unbound" or "off-ground" projects managed by non-Alsatian companies with no cultural connections or history in those territories.

The most recent, ongoing study dealing with public perceptions of deep geothermal energy in Alsace is related to participatory science. This can be defined as forms of scientific knowledge production in which civil society actively participates. Such projects promote dialog between science and society. For instance, based on a scientific approach, scientists contribute to exchanges with citizens on growing concerns about induced seismicity related to deep geothermal energy in Alsace (https://anr.fr/Projet-ANR-21-CE05-0033). The University of Strasbourg has launched a new research project that involves an innovative way of testing a new collaborative geohazard monitoring paradigm in an urban context. The basic principle is to deploy a large number of cheap seismological sensors, working closely with mining authorities and citizens to get a dense seismological dataset. This research project was proposed by scientists after the seismic events which took place in Strasbourg in December 2020. It aims at evaluating the seismic risks induced by the geothermal operations and how they are perceived by the population. The results will be co-analyzed by scientists and non-seismological experts. Based on such a collaborative approach, we can measure public involvement and how induced seismicity related to industrial operations is perceived and represented in the URG.

On the French side of the URG, local populations are also invited to participate in public inquiries when geothermal projects are close to the operational stage. Organized by the local Prefecture, public consultations consist in making available technical documents that are in most cases difficult for citizens not familiar with science and technology to understand [20]. These consultations are seen as a way of measuring acceptability more than active participation by citizens. For instance, low participation by the public in such inquiries is interpreted as a high level of acceptance of the project (low level of protest). The main criticism voiced by inhabitants concerns the potential impacts of drilling operations on their environment, such as induced seismicity, groundwater pollution and radioactivity issues [20]. Moreover, the technology used to produce electricity (Organic Rankine Cycle) is also liable to raise some major concerns. The risk of explosion due to the presence of isobutane in the power plant was a major argument against one project in the Strasbourg area, where the site had pre-existing industrial risks (Seveso zone). A consensus between inhabitants, residents' associations and local politicians against this urban geothermal project led the operator to give up on this site.

3. A look-back at the evolution of seismic monitoring

As mentioned in the previous section, it has been shown that stimulation operations (even moderate ones) and exploitation of the geothermal loop are likely to generate induced micro-seismic activity, temporarily or continuously [15, 21], which can occasionally be felt by local population [22, 23]. In extreme cases, the occurrence of induced seismicity may lead geothermal operations to be shut down, as was the case of the geothermal project in Basel [24] or Vendenheim [11]. As a result, it became common in deep geothermal energy projects, which target naturally fractured reservoirs, to monitor geothermal activities with high-sensitivity through seismological networks. Initially, on the European EGS pilot site at Soultz-sous-Forêts, the seismic monitoring was dedicated to understand the fracture network activated by hydraulic stimulation, and in order to get an image of the reservoir development. More recently, the objective of the seismic monitoring then switched to the necessity to minimize the seismic risk [25].

These networks are generally designed to accurately assess the state of the natural seismicity before any operation, but also to detect any emergence of induced micro-seismic activity attributable to the geothermal operations.

Until 2015, there was no regulatory framework in France to supervise the environmental monitoring of geothermal plants. The deployment of such monitoring networks was left up to the operators. The experience acquired at the Soultz-sous-Forêts and Rittershoffen geothermal plants, together with the growing number of operators involved in deep geothermal energy, highlighted the need to monitor these systems. For this purpose, the French mining authorities established clear rules to regulate the construction, development and exploitation of geothermal plants. The same happened in Germany, where the mining authorities of regions (e.g., the Palatinate) developed their own regulatory framework. In Switzerland, the canton recommends good practices but does not impose a clear regulatory framework.

Table 1 gives a comparison of the different regulation on seismic monitoring in France, Germany and Switzerland.

Along with this clarification of the regulations, the operators have developed their workflows to clarify the decision-making process in event of defined thresholds being exceeded (see **Figure 3**). Up to now, by following this procedure French operator Électricité de Strasbourg has managed to keep the seismic risk under control since no induced seismicity has been felt by local population around the exploitation of Soultz and Rittershoffen geothermal plants since 2016 or during the drilling of GIL-1 well in Illkirch [16].

In the future, it can be expected that new ways of managing seismic risk on site will appear, with the use of predictive tools to anticipate the occurrence of events that could be felt by the population [26].

	France/Alsace	**Germany/Palatinate**	**Swiss**
National or regional framework	Regional	Regional	Cantonal
Authority	Local mining authority	Local mining authority	Local mining authority
Regulation	Prefectural decree (legifrance.gouv.fr)	DIN 4150 (beuth.de/de/norm/din-4150-3)	ETH Good Practice Guide [26]
Seismic network Mandatory	Yes	Yes	Recommended, category III for traffic light system

	France/Alsace	Germany/Palatinate	Swiss
Seismic network technical requirement	• Real-time data monitoring • 4 short period velocimeters • 1 "multi-sensor" station including a broad-band seismometer, an accelerometer, a GNSS receiver and a corner-coin reflector	• One network designed with four velocimeters according to the DIN 4150 (on buildings) • One real-time network, that can use both accelerometers or velocimeters; • Number (generally 4 stations) and kind of stations of this network needs to be coordinated with the mining authority	• Real-time data monitoring • Shallow borehole (80–150 m depth) • At least 3 stations around the project • One station in the center of the network equipped with an accelerometer • 3 orthogonal components for all stations
Seismic network installation	6 months before start of drilling	3 months before or at the latest when drilling starts	6 months before stimulation
Public data	All data from the "multisensory" station	All data from the velocimeters used to monitor DIN 4150 compliance	Possibly all data
Organization collecting public data	ReNaSS (French seismic monitoring network)	The local (normally state operated) seismic observation network	SED (Swiss Seismological Service)
Definition of threshold	Yes	Yes	Yes
Physical value used for defining thresholds	PGV	PGV	ML and PGV
Number of thresholds	3 thresholds (measured on 2 stations): • 0.5 mm/s, close monitoring • 1.0 mm/s, short term reduction in flow • 1.5 mm/s, stop operation	5 thresholds (measured on 1 station only): • 0.2 mm/s, daily reporting • 0.5 mm/s, short term reduction in flow • 1 mm/s, long term reduction in flow • 5 mm/s, operate at minimum flow • 10 mm/s, stop operation	To be defined with local mining authority
Reporting to the mining authority	Before drilling: report on natural seismicity Drilling: monthly Testing: daily Operation: monthly	Before drilling: report on natural seismicity and a seismic hazard analysis Testing: daily Operation: monthly	To be defined with local mining authority

Table 1.
Comparison of the regulation on seismic monitoring in France, Germany and Switzerland (regulations in force in 2021).

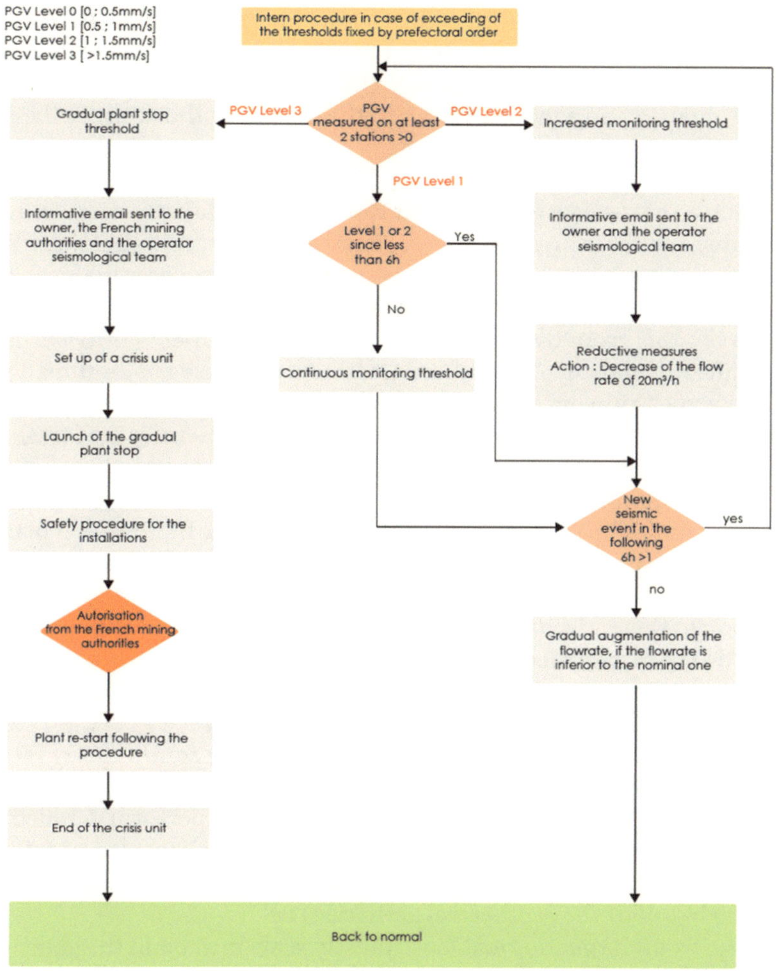

Figure 3.
Decisional chart designed by Électricité de Strasbourg in case of occurrence of induced micro-seismic activity, based on French mining authority regulation.

4. Environmentally friendly energy source

4.1 Environmental impacts

Apart from induced seismicity, which is presented in the previous section of this chapter, geothermal energy can induce other disturbances in the environment and for residents. Five kinds of impacts are identified, minimized when possible and monitored [27]. They correspond to slow surface deformations, shallow groundwater protection, contamination of soils by geothermal brine leakages, precipitation of radioactive scales in surface infrastructures, and noise from the geothermal plant. Their main impacts and mitigation schemes are presented below:

Slow surface deformations (subsidence, uplift) has been identified as a potential major impact of the plants on their surrounding areas since it was reported for the Landau power plant in the German part of the URG [14]. To identify such slow vertical ground motion before it reaches significant deformation, a telemetered GNSS

(Global Navigation Satellite System) receiver is installed on each geothermal plant platform. Additionally, one corner reflector is installed to measure surface deformations through satellite radar interferometry (InSAR technique). Results of the GNSS monitoring is reported on a monthly basis to the mining authorities. To date (July 2022), no significant ground motion has been observed at the different geothermal sites in the French part of the URG.

The protection of shallow groundwater resources is a major concern in areas where the Rhine aquifer is present. This regional aquifer represents 35,000 millions of cubic meters, making it one of the biggest freshwater resources in Europe. It is extremely important for the region's economic development and its drinking water supply. However, this groundwater resource is very vulnerable. More than a third of its surface is already undrinkable without treatment, due to various human activities (https://www.ermes-rhin.eu/FR/documents-et-publications/copy-of-acc%C3%A8s-libre.html). Since 2014 new projects have been in development in the Strasbourg area (Illkirch, Vendenheim), where this aquifer can reach a thickness of 150 m. For these projects, the design of the geothermal wells involves isolating the geothermal water from the aquifer by three cemented casings. Over the lifetime of a plant, the casings should be inspected on a 3-year basis for the injection well and 6-year basis for a production well. All these inspections must be reported to the mining authorities. In addition to these mechanical barriers, a piezometric monitoring network has been deployed. The monitoring starts 3 months prior to the drilling and continues after the drilling and well testing phases. For instance, during all the geothermal activities at the Illkirch site, physico-chemical parameters such as pH, Eh (redox potential), conductivity, temperature and water level were continuously monitored and were available remotely in real time, to quickly detect any possible leakage. The Rhine water was sampled before and during the drilling to perform detailed chemical analyses and monitor possible contamination due to the geothermal activities. In Illkirch, an important result was that the Rhine aquifer water remained drinkable and unpolluted during all of the geothermal activities [16].

The leakage from the geothermal loop, mostly scale formed in the plant's piping and geothermal water could lead to the contamination of soil or surface water in the vicinity of the geothermal plant. Indeed, geothermal water is highly saline (over 3 times more than seawater [28]) and the scales are currently mostly made of galena (PbS) which contains heavy metal and radionuclides [29]. This risk is managed at different levels. A fundamental parameter to assess, to prevent leakage, is corrosion in the geothermal loop. Corrosion is a major issue that must be taken into account in plant design. To prevent corrosion issues, the most strategic parts of the geothermal loop (heat exchanger, filters, some valves and pumps) are made of a specific grade of steel which shows high resistance to corrosion from the URG geothermal water [30]. Less critical parts are usually made of carbon steel including an over thickness to ensure their longevity. During the operation, corrosion inhibitors are used and corrosion monitoring is performed through either coupon testing or corrosion probe. Additionally, several parts of the geothermal loop are inspected on a regular basis to check the current corrosion situation and anticipate part replacement if needed. In the design of the plant, a specific water management system is set up to keep the different types of water (geothermal water, wastewater and rainwater) apart and prevent the geothermal water contaminating the environment. The rainwater at the power plant is collected in a tank by gravity and treated with a sand filter together with a scrubber and an oil separator. Before releasing the rainwater into the environment, an operator checks its conductivity. If this measurement indicates contamination of the rainwater by the geothermal water, it is pumped to the geothermal storage pool and then reinjected into the geothermal reservoir.

As mentioned above, scale formed from the geothermal water can contain radionuclides [31], mostly ^{226}Ra and ^{210}Pb and their respective daughter elements. Therefore, it is important to keep the dose rate as low as reasonably acceptable for both, the workers and the public. Periodic radioactivity and dose rate measurements are performed on site to evaluate the risk within the installation, identifying zones with restricted access (where the public is not allowed) and where wearing a dosimeter is mandatory. In addition to these monitoring, an important research and development effort has been made since 2009 to reduce mineral precipitation in the power plant. This project led to the elimination of barite which was the radium bearing mineral [32, 33]. Whereas ^{226}Ra has a long half-life and generates important gamma-ray emissions that can propagate trough the pipes, ^{210}Pb is a short half-life radionuclide (less than 30 years) emitting mostly alpha and beta rays that cannot propagate through the pipes. Thus, the main remaining risk is currently inhalation and ingestion. Since 2020, an annual dust and radon measurement campaign has been performed to assess this risk for workers and in the area around the plants. These measurements are not significantly higher than the reference points outside of the plant [34].

Noise near the geothermal plant has also an impact on local residents. Indeed, in 2012 when an acceptability survey of the Soultz power plant was performed, collecting the perceptions of the inhabitants of Soultz-sous-Forêts and Kutzenhausen, the most cited impact was noise from the plant [18]. However, at that time, the air cooling system had a defect, which was temporarily generating an abnormal level of noise, but this was quickly corrected.

In France, geothermal plants must meet maximum noise emission levels at the boundary of the facility, i.e. noise must not exceed 70 dB(A) during the day and 60 dB(A) at night, except if the ambient noise is above these limits. Additionally, it should not increase the ambient noise in surrounding regulated noise area—such as a residential area—more than the values presented in **Table 2**.

During the plant design phase, a noise impact study is performed for the selection of low-noise emission equipment, such as the air condenser, but also for proper positioning of the equipment on the power plant platform. It can provide recommendations for sound insulation (for instance, anti-noise wall around the plant) to respect the noise regulations.

4.2 Life cycle assessments

As a complement to the site-specific analysis of environmental impacts and their respective mitigation schemes, on a general level, the environmental impacts of

Ambient noise in regulated area, including noise emission of the project	Acceptable emergence values between 7 am and 10 pm, excepted Sundays or public holidays	Acceptable emergence values between 10 pm and 7 am, and on Sundays or public holidays
Between 35 dB(A) and 45 dB(A)	6 dB(A)	4 dB(A)
Above 45 dB(A)	5 dB(A)	3 dB(A)

Table 2.
Acceptable sound emergence values in the regulated areas of the geothermal projects in northern Alsace.

geothermal energy can be assessed using the Life Cycle Assessment (LCA) methodology. LCA is a widely accepted and standardized methodology which translates the resources, materials and energy flows necessary for the entire life cycle of a system into a series of potential environmental impacts [35, 36]. To make sure that geothermal energy does not involve additional environmental impacts, LCA can provide very valuable information to ease decision-making processes and make comparisons with other energy sources, even if some potentially relevant environmental impacts are still missing from current LCA methodology, such as noise or seismicity.

The first LCA of a geothermal plant in the URG was published Lacirignola et al. in 2013 [37]. This comprehensive study presents the environmental performances per kWh of electricity of the Soultz geothermal plant considering different design options. Greenhouse gas (GHG) emissions contributing to global warming were estimated at about 36.7 gCO_2eq/kW. The GHG emissions from power generation of the Soultz geothermal plant appeared to be lower than the average GHG emissions of the French electricity mix, which are about 58 gCO_2eq/kW or German electricity, 349 gCO_2eq/kW (https://www.statista.com/statistics/1291750/carbon-intensity-power-sector-eu-country/). This LCA was then used to propose a simplified model for the estimation of greenhouse gases emitted by an enhanced geothermal system for power generation [38].

The GHG emissions of the Rittershoffen geothermal heat plant were first assessed by [39]. This preliminary work was then completed in line with the methodological guidelines developed as part of this European H2020 GEOENVI research project

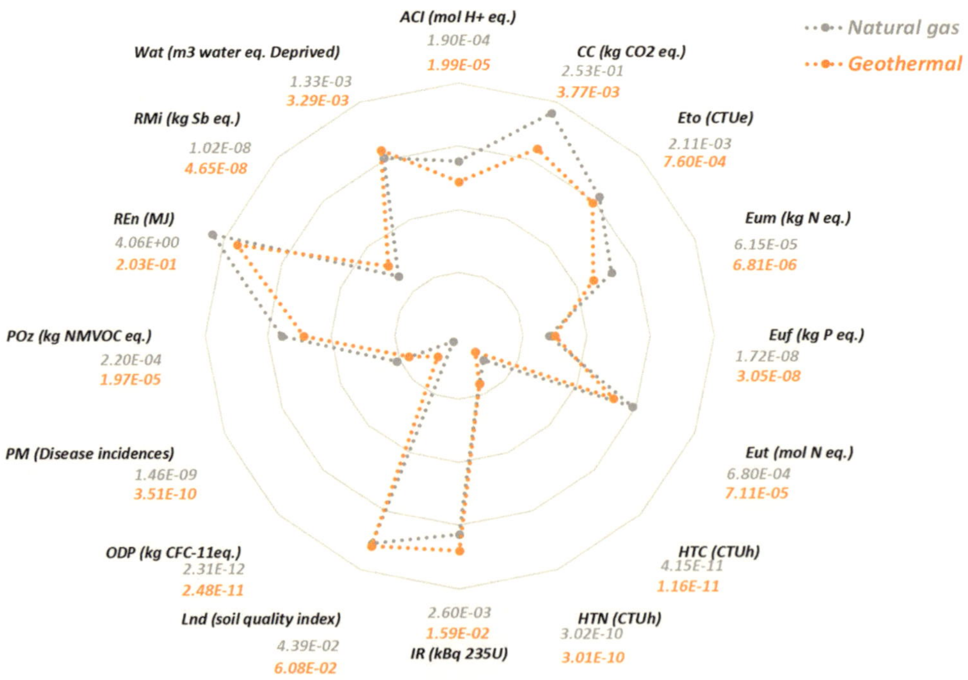

Figure 4.
Comparison between the production of 1 kWh of heat from natural gas and from the Rittershoffen geothermal plant for different impact categories: Acidification (ACI), climate change (CC), freshwater ecotoxicity (Eto), marine eutrophication (Eum), freshwater eutrophication (Euf), terrestrial eutrophication (Eut), human toxicity-cancer (HTC), human toxicity-non-cancer (HTN), ionizing radiation-human health (IR), land use (Lnd), ozone depletion (ODP), particulate matter (PM), photochemical ozone formation-human health (POz), resource use-fossils (REn), resource use-minerals and metals (RMi), and water use (wat). From ref. [41].

which provided recommendations to harmonize methodological choices in each of the four steps of LCA [40]. Environmental performances per kWh of heat of the Rittershoffen geothermal plant were published by [41] and compared to at the production of 1kWh of heat with natural gas. Results of this study are presented in **Figure 4**. A parameterized model for enhanced geothermal system for heat production was established based on the Rittershoffen geothermal plant LCA to assess seven environmental criteria [42].

According to **Figure 4**, the potential impacts of the Rittershoffen geothermal heat plant are similar to or lower than those of natural gas in most impact categories. In particular, the potential impact on Climate change is estimated at 3.7 gCO_2eq/kWh for the Rittershoffen geothermal heat plant. This impact is 67 times lower than that of heat from natural gas. The only exception is the Human Health—Ionizing radiation impact category, for which the environmental impact is higher for the Rittershoffen geothermal plant. This impact is indirect, due to the electricity consumption during operation of the geothermal heat plant provided by the French electricity mix, which is 75% nuclear. A mix with a higher share of renewable energy, or a heat plant with on-site self-power generation, would automatically reduce the impact in this category.

The Soultz and Rittershoffen LCAs confirm that geothermal energy generated under URG conditions has very low environmental impacts. This energy appears to be a promising renewable energy source for the decarbonization of power generation, district heating and heat used in industrial processes in Europe.

5. Contribution of geothermal energy to local sustainable development

5.1 Impact on the local economy

The impact of deep geothermal energy on the local economy is unfortunately not well documented. This impact can be assessed using Life Cycle Cost Analysis (LCCA). LLCA is a tool mainly used to determine the most cost-effective option for an object or process among different alternatives and over its entire lifetime. LLCA also provides information about the origin of purchase, supplier typology or the creation of added value. That information can be used to assess the impact on the local economy.

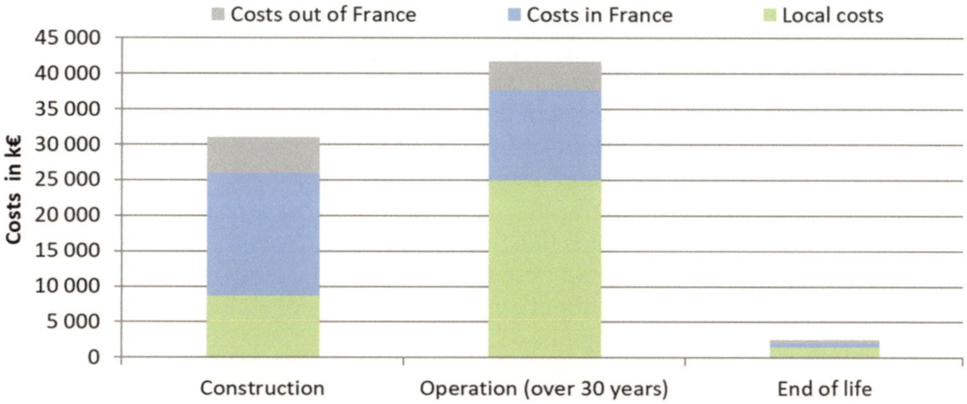

Figure 5.
Life cycle costs of the Rittershoffen geothermal plant.

Perez et al. published a first study assessing the environmental and economic impact of the Rittershoffen geothermal plant [43]. An LLCA was performed for the entire project lifetime. The life cycle costs of the Rittershoffen geothermal plant (**Figure 5**) were detailed for different levels (local, i.e., Department of Bas-Rhin, national, i.e., France, and international) and for 4 project stages: (1) Exploration and drilling, (2) Geothermal plant construction, (3) Geothermal plant operation, and (4) End of life.

This first study clearly confirmed the benefits of the deep geothermal project in the URG for the local economy. Indeed, about 45% of expenditure over the lifetime of the project benefits the local economy and about 87% the national economy. The construction, operation and end-of-life phases have stronger relative impacts on the local economy: respectively 48, 60, and 57% of the total costs compared to exploration and drilling, which only contributes 19% of the total costs. Indeed, drilling and service companies are mostly part of the upstream oil and gas industry, which is located outside the department of Bas-Rhin or even outside France.

5.2 Impact on local employment

Impacts on the local economy can also be evaluated according to the impact on local employment. The study in [43] also assessed this aspect, based on the Rittershoffen geothermal plant LCCA. In this study, direct employment was defined as workers who are employed by companies involved in the different stages of the life cycle of the Rittershoffen project and indirect employment as job creation in the local economy due to demand created by the project and its direct employees. Indirect employment is unfortunately very difficult to assess, and as a result this study focused on direct employment within the boundaries of France. The unit used in this employment assessment study is full-time equivalents (= 1607 h/year) (FTEs).

Input data used for this study were the plant owner's accounting and other data on its suppliers such as legal structure, location, business identification, activity sector and economic data (turnover, added value and average number of permanent staff). These data were enriched with national economic statistics extracted from data produced by INSEE (National Institute of Statistics and Economic Studies), which collects, analyses and disseminates information on the French economy and society. Associating costs with the location of the companies involved in the life cycle costs of the Rittershoffen geothermal plant has made it possible to identify where the direct jobs are located: either at local level in the French Department of Bas-Rhin or at national level (**Figure 6**).

Exploration, drilling and plant construction, from early 2012 to mid-2016, occupied about 124 FTEs. Annually, maintenance and operation of the geothermal plant has involved about 4 local FTEs and 1 in the rest of France, which amounts to about FTEs over 25 years of operation. End-of-life jobs are estimated at about 12 FTEs over a period of 4 to 6 months for well cementing and plant dismantling.

This study shows that the geographical distribution of the direct employment within France during the life cycle is like that of the life cycle costs. Direct employment is more important at local level than at national level during the operation and construction phases. Conversely, direct employment is stronger at national level during the exploration and drilling phases. It is the local economy that has mainly benefited from direct employment resulting from the Rittershoffen project, with 63% over the entire lifetime of the project and 82% during the 25 years of operation.

Further investigations are nevertheless required to assess the impact on local indirect employment.

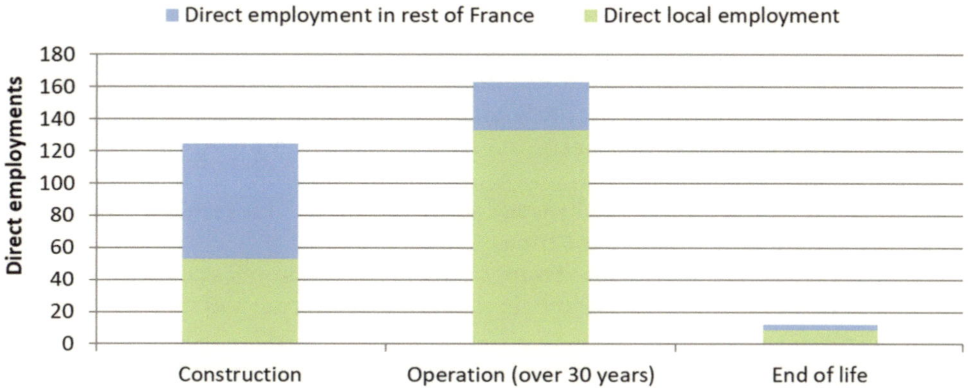

Figure 6.
Direct employment in France during the life-cycle of the geothermal plant at Rittershoffen.

5.3 Impact on local attractivity

Geothermal energy is a local, non-intermittent, renewable and decarbonized energy source. In the context of global warming, the taxonomy and the high variability of natural gas, this energy source is an opportunity for the economic attractivity of a region, especially in terms of heat production. Indeed, according to a study published by ADEME, the French agency for the ecological transition, in 2020, deep geothermal and waste heat are the most economical energy sources for industry and the residential sector [44]. The cost of heat from deep geothermal energy was in a range of €15 to 55/MWh, while the cost of heat from natural gas was €51–85/MWh [44] before the Ukrainian crisis and the rise of the natural gas price. Deep geothermal energy can really boost the economic development of a territory by attracting energy consumers or reducing the carbon footprint of existing facility.

Since 2019, over 60% of the heating needs of Bruchsal police headquarters have been supplied by the nearby geothermal power plant. The reduction in GHG emissions has reached 700 ton/year. Thus, the Bruchsal geothermal power plant demonstrates that in the URG geothermal energy can supply a climate-friendly alternative to fossil fuel heating locally, safely and reliably.

6. Towards additional valorization to increase local benefits

6.1 Geothermal power production and use of the residual heat in the URG

The deep geothermal power plants in the URG produce saline water at temperatures above 150°C by pumping up fluid from deep geothermal reservoirs. The hot fluid produced is used to generate electricity by means of an Organic Rankine Cycle (ORC) or to supply heat to industrial users, following which the brine is reinjected back into the fractured granite reservoir at around 70°C. Reinjection at a lower temperature was deemed non-feasible due to the build-up of scale in the heat exchanger below 60°C. However, recent studies carried out as part of the European MEET project suggest that no new scaling was found at lower temperatures at Soultz [29, 45, 46] and that operators could reinject at 40°C, valorizing 20–45% of residual thermal energy.

Thus, a generic research study has been conducted to identify low-temperature industrial processes that could use the residual heat produced from ORC-based power plants in this region [47]. As the ORC cooling system is usually using ambient air, an energy analysis was carried out for the Soultz-sous-Forêts power plant to establish a relationship for the residual thermal power with the outside air temperature. Based on that relationship, a residual thermal energy profile was produced for a theoretical ORC power plant with a brine production rate of 200 m^3/h. This resulted in a thermal power of 6–8 MW depending on the time of year.

Therefore, there are various low-temperature industrial applications that could harness this residual heat to reduce the reinjection temperature and increase economic and energy efficiency.

Based on the residual thermal power available and a maximum temperature constraint of 70°C, extensive market research has been conducted to identify the industrial processes that might use low-temperature heat [47]. Several aspects were taken into consideration such as the application's current availability, projected development, availability of companies, environmental constraints, distribution of industrial units and local actors. The operators of the activities and applications identified were consulted to identify the most promising processes for residual heat valorization, and an implantation study and economic assessment were carried out in order to determine the levelized cost of geothermal heating (LCOH). The main applications are drying processes (sludge from wastewater treatment, brewery waste recovery, wood sector), aquaculture (fish farming, spirulina production, aquaponics), agriculture, insect protein production, biogas production and district heating. The various applications considered in the market study are outlined in **Figure 7**. The applications painted in red were explored further to determine their feasibility in the techno-economic evaluation [47]. The results of the techno-economic study provided key insight for geothermal companies with regard to the applications that can be used to valorize residual heat and the production capacity required for economic feasibility.

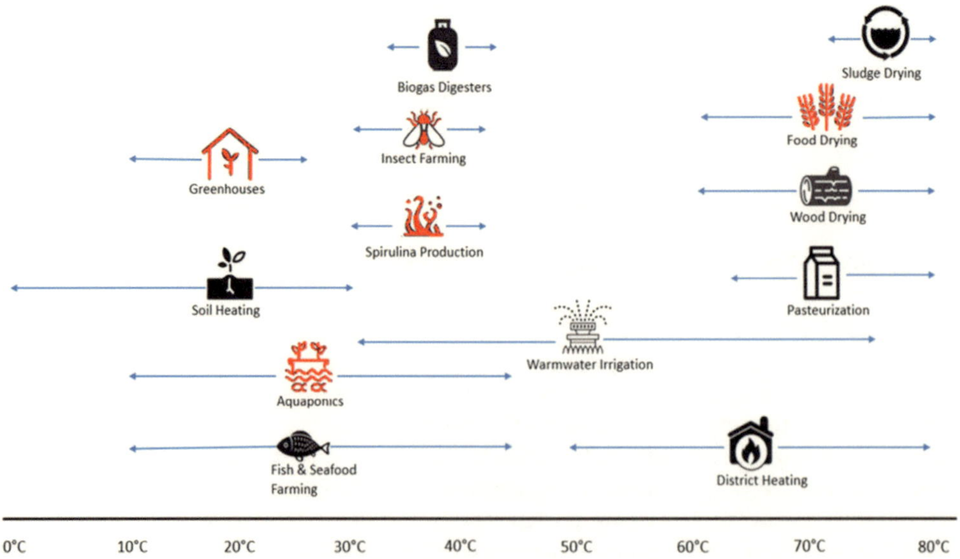

Figure 7.
Temperature ranges for various industrial applications.

In conclusion, geothermal energy can significantly improve the sustainability of the food sector by providing heat to new innovative technologies such as insect protein production and farming methods such as aquaponics, greenhouses, fish farming and spirulina production. Geothermal power plants can really increase local benefits by supplying the residual heat at low to marginal cost, creating an economic ecosystem in the surrounding area and indirect jobs.

6.2 Geothermal power production, critical raw materials, and lithium extraction in the URG

Harvesting geothermal power (heat and electricity) from subsurface reservoirs is a process that can be used as a renewable energy source by the local population and industries, thanks to the geothermal power plants. However, profits associated with a geothermal power can be hard to maintain (heat and electricity sales), and strategies must be developed to ensure profitable growth. To improve the economics of geothermal power plants, numerous investigations have been carried out to find ways of coupling the production of geothermal energy with the extraction of minerals and metals dissolved in the fluid [48, 49]. Numerous chemical elements in these dissolved solids accumulated in the solution due to weathering of the rocks are a potential source of valuable metals and minerals: critical raw materials (CRMs). CRMs are defined as raw materials which are economically and strategically important for the world economy, but which have a high risk associated with their supply [50]. They are essential to the functioning of a wide range of industrial and public activities (consumer electronics, health, steelmaking, space exploration, etc.). It is well known that such CRMs are contained in geothermal water, and recovery methods are developed where the process is considered to be economically profitable now or in the future [49]. Silica, zinc, lithium, manganese, potassium and a number of rare earth elements are among the most studied elements due to their high concentrations in geothermal fluids [51, 52]. Although CRMs concentrations in the geothermal fluids are lower than what is measured in mineral ores (e.g., ppm in brines vs. % in ores for lithium [53, 54] the costs associated with CRMs extraction are potentially low for the following reasons set out by [49]:

The associated costs will be divided between power and mineral production. Mineral extraction would be developed at geothermal power plants that already exist where the technical staff on site already have a good knowledge of the surface installations.

There are no costs associated with the beneficiation of the minerals/metals, which normally include disaggregation, physical separation (gravity and magnetic separation), and chemical separation (leaching, froth flotation).

In spite of a lower concentration than in ores, the geothermal process involves large quantities of water (e.g., around 300 m^3/h produced at Rittershoffen power plant) and therefore a high quantity of CRMs could be extracted.

Concentrations of these metals in the fluid depend on several parameters that affect the chemical composition of the water during its underground circulation: 1. Chemical composition and properties of the rocks (e.g., mineral content and porosity); 2. Initial composition of the fluid (e.g., rainwater, pH); 3. Duration of interaction with the rocks; 4. Temperature and pressure during fluid/rock interaction; 5. Fluid/rock ratio in terms of volume. 6. Possible anthropogenic influence. Geothermal fluid compositions are therefore dependent on their location in the world (e.g. subsurface geology), and metal recovery technologies must be developed to adapt to the different properties of the fluids. In the URG, these fluids have high concentrations of dissolved solids (~100 g/L), mostly Cl, Ca, and Na [28, 34, 55] due to the water movement through the different in-situ geological units (sedimentary and granitic rocks).

In the URG, it is possible to identify several CRMs with interesting concentrations that can be profitable in the future [34]. **Table 3** summarizes the valuable CRMs found in Rittershoffen and Soultz-sous-Forêts geothermal fluids.

Possible annual income from CRM extraction is closely linked to the quantities of lithium in the geothermal fluids (about ~88% of the total income at

Geothermal power plant	CRMs	Mean concentration in the brine (mg/L)	Market interest	Prices in 2022 (€/ton)	Possible income per year (€)
Soultz-sous-Forêts	Mg	136	Magnesium powder, Mg alloy and Mg ingot	5000	5.42E+05
	Sr	431	Strontium metal	14,135	4.84E+06
	Li	168	Lithium carbonate, Li hydroxide monohydrate, Li chloride	60,000 for LCE	4.27E+07
	Si	89	Silicon metal, Si powder	3000	2.12E+05
	Zn	2.7	Zinc ingot, Zn oxide, Zn powder	3600	7.81E+03
			Possible Income (€/L)		0.044
			Possible income (€/year)		4.83E+07
Rittershoffen	Mg	123	Magnesium powder, Mg alloy and Mg ingot	5000	1.05E+06
	Sr	459	Strontium metal	14,135	1.11E+07
	Li	181	Lithium carbonate, Li hydroxide monohydrate, Li chloride	60,000 for LCE	9.86E+07
	Si	98	Silicon metal, Si powder	3000	5.00E+05
	Zn	3.3	Zinc ingot, Zn oxide, Zn powder	3600	2.04E+04
			Possible Income (€/L)		0.047
			Possible income (€/year)		1.11E+08

Source of average CRM prices: https://ise-metal-quotes.com, July 2022.

Table 3.
Possible income associated with the extraction of CRMs at Rittershoffen and Soultz-sous-Forêts (heat and electricity sales not included). The flow rate at Rittershoffen is 75 L/s and 35 L/s at Soultz-sous-Forêts. For the calculation, a mineral extraction of 80% was assumed, and a plant availability of 90%.

Soultz-sous-Forêts and Rittershoffen, **Table 3**). Lithium is a strategic metal, especially for electric vehicle battery manufacturing, for which worldwide demand is constantly increasing. Analysts forecast lithium demand approaching 1 Mt. LCE (Lithium Carbonate Equivalent, Li_2CO_3) by 2026 [53, 56]. Although lithium is found ubiquitously in the environment, the URG geothermal waters are rich in lithium with an average concentration measured between 150 and 210 mg/L [28, 34, 55]. This significant Li concentration (~1000 times more concentrated than in sea water) is therefore of great interest for future exploitation. For instance, one doublet at Rittershoffen (one production and one injection well) could produce more than 1500 tons of LCE per year assuming a plant availability of 90% and a mineral extraction yield of 80%. Given that current worldwide lithium production is concentrated in Australia, Chile, Argentina and China, the production of lithium at geothermal power plants should help the European Union (EU) to reduce its dependency on other countries and to produce more sustainable lithium [53]. Additionally, the operation of a power plant with two doublets producing ~3000 tons of LCE, would create new jobs. In total 72 employees would need to be hired for the effective operation of the lithium production plant including maintenance and lab teams, an operations team, managers and additional staff to cover for holidays and absence.

As part of the EuGeLi (European Geothermal Lithium Brine) project, a collaborative research and innovation project launched in January 2019, direct lithium extraction (DLE) tests were conducted with real brine in geothermal exploitation conditions (80°C and 20 bar) on site. The work managed to recover several kilograms of precipitated battery-grade Li_2CO_3, showing the feasibility of directly extracting Li from geothermal fluids. However, several parameters need to be adjusted to improve and increase the productivity of the DLE process and the overall recovery level. Adjustment of the flow rate in the column, comprehension of the chemical reaction occurring between the brine and the active solid over the long term and management of higher impurities in lithium solutions are among the main parameters to consider in the future to improve the profitability of the project [57]. Given the novelty of the lithium extraction process and the lack of long-term, large-scale operational experience, numerous factors could influence the long-term profitability of DLE such as: a decline over time of the lithium concentration in the production fluids due to poor re-saturation of lithium in the brine after extraction; a drop in lithium prices due to technological shifts or alternative technologies affecting demand for lithium; poor acceptability resulting in less public support through subsidies or permit approvals [58].

Although DLE is a young technology, it should significantly reduce carbon emissions compared to other methods of producing and refining lithium (hard rock mining and evaporation ponds). According to Vulcan Energy and the project Zero Carbon LithiumTM, one ton of LiOH (lithium hydroxide) produced by hard rock mining emits 15,000 kg of CO_2 compared to zero for harvesting geothermal lithium (https://v-er.eu/zero-carbon-lithium/, July 2022).

7. Conclusion

Despite drawbacks due to several felt induced seismic events, deep geothermal energy is still being developed in the Upper Rhine Graben and has a promising future. The operators and the mining authorities have introduced best practices and appropriate rules to mitigate the seismic risk and to minimize other environmental impacts such as slow surface deformation, shallow groundwater protection, contamination of soils by geothermal brine leakages, radioactivity and noise. Life Cycle Assessments highlight the overall low environmental impact of this source of energy. In particular, the potential impact on climate change is estimated at 3.7 gCO_2eq/kWh for the Rittershoffen geothermal heat plant. This impact is 67 times lower than that of heat from natural gas. In terms of socio-economic impact, a Life Cycle Costs Analysis performed on the Rittershoffen plant showed how this industry is well anchored in its territory with around 60% of its costs benefiting local actors during the operating phase. Nevertheless, deep geothermal energy in the URG is still in the early stages of its development and additional valorization such low temperature heating and/or metal extraction from the geothermal water could help the deployment of this industry on a larger scale. This would increase the attractivity of the region by providing heat for agro-industries at reasonable prices and lithium which could be a starting point to develop a new industrial sector in the area.

Acknowledgements

The authors would like to thank several projects whose results have been used in this chapter. The European DESTRESS, GeoEnvi and MEET European projects are mostly funded by the European Commission's Horizon 2020 Research and Innovation Program under grant agreements No 691728, No 818242, and No 792037 respectively. Some results were partly funded by ADEME (the French Agency for Ecological Transition) as part of the Eranet Geothermica Zodrex project and the EGS Alsace project co-funded by Electricité de Strasbourg. The authors acknowledge the GEIE EMC and ECOGI for providing the data. The authors also acknowledge the EuGeLi project (European Geothermal Lithium brines), co-funded by the European Union, under the European Institute of Technology (EIT) Raw Materials Program.

Conflict of interest

The authors acknowledge that they are employed by ES-Geothermie which is a developer and operator of deep geothermal plants in France.

Author details

Eléonore Dalmais[1]*, Guillaume Ravier[1], Vincent Maurer[1], David Fries[1], Albert Genter[1] and Béatrice Pandélis[1,2]

1 ES-Géothermie, Strasbourg, France

2 Electricité de Strasbourg, Strasbourg, France

*Address all correspondence to: eleonore.dalmais@es.fr

References

[1] Vidal J, Genter A. Overview of naturally permeable fractured reservoirs in the central and southern upper Rhine graben: Insights from geothermal wells. Geothermics. 2018;**74**:57-73

[2] Dalmais E, Genter A, Reinecker J, Pandélis B. La géothermie profonde dans le Fossé rhénan supérieur, des années 80 à aujourd'hui. Géologues Géosciences et Société. 2021;**210**:83-91

[3] Vidal J, Genter A, Schmittbuhl J. How do permeable fractures in the Triassic sediments of northern Alsace characterize the top of hydrothermal convective cells? Evidence from Soultz geothermal boreholes (France). Geothermal Energy. 2015;**3**(1):1-28

[4] Baujard C, Genter A, Dalmais E, Maurer V, Hehn R, Rosillette R, et al. Hydrothermal characterization of wells GRT-1 and GRT-2 in Rittershoffen, France: Implications on the understanding of natural flow systems in the Rhine graben. Geothermics. 2017;**65**:255-268

[5] Evans KF, Moriya H, Niitsuma H, Jones RH, Phillips WS, Genter A, et al. Microseismicity and permeability enhancement of hydrogeologic structures during massive fluid injections into granite at 3 km depth at the Soultz HDR site: Induced seismicity and flow in deep granite. Geophysical Journal International. 2004;**160**(1):389-412

[6] Dempsey D, Kelkar S, Davatzes N, Hickman S, Moos D, Zemach E. Evaluating the roles of thermoelastic and poroelastic stress changes in the desert peak EGS stimulation. In: Proceedings, Thirty-Ninth Workshop on Geothermal Reservoir Engineering. Stanford, California: Stanford University; 2014. p. 15

[7] Schill E, Genter A, Cuenot N, Kohl T. Hydraulic performance history at the Soultz EGS reservoirs from stimulation and long-term circulation tests. Geothermics. 2017;**70**:110-124

[8] Charléty J, Cuenot N, Dorbath C, Dorbath L. Tomographic study of the seismic velocity at the Soultz-sous-Forêts EGS/HDR site. Geothermics. 2006;**35**(5-6):532-543

[9] Trutnevyte E, Azevedo IL. Induced seismicity hazard and risk by enhanced geothermal systems: An expert elicitation approach. Environmental Research Letters. 2018;**13**(3):1-9

[10] Häring MO, Schanz U, Ladner F, Dyer BC. Characterisation of the Basel 1 enhanced geothermal system. Geothermics. 2008;**37**(5):469-495

[11] Schmittbuhl J, Lambotte S, Lengliné O, Grunberg M, Jund H, Vergne J, et al. Induced and triggered seismicity below the city of Strasbourg, France from November 2019 to January 2021. Comptes Rendus. Géoscience. 2021;**353**(S1):561-584

[12] Terrier M, De Santis F, Soliva R, Valley B, Bruel D, Geraud Y, et al. Rapport Phase 1 du comité d'experts créé en appui à l'administration sur la boucle géothermique GEOVEN. Open File Report, Préfecture du Bas-Rhin. 2022. Available from: https://www.bas-rhin.gouv.fr/Politiques-publiques/Environnement/Geothermie/Rapport-du-comite-d-experts-cree-en-appui-a-l-administration-sur-la-boucle-geothermique-GEOVEN

[13] Schindler M, Baumgärtner J, Gandy T, Hauffe P, Hettkamp T, Menzel H, et al. Successful hydraulic stimulation techniques for electric

power production in the upper Rhine graben, Central Europe. In: Proceedings World Geothermal Congress 2010. Bali, Indonesia: International Geothermal Association; 2010. p. 7

[14] Heimlich C, Gourmelen N, Masson F, Schmittbuhl J, Kim S-W, Azzola J. Uplift around the geothermal power plant of Landau (Germany) as observed by InSAR monitoring. Geothermal Energy. 2015;**3**(1):2

[15] Maurer V, Gaucher E Grunberg M, Koepke R, Pestourie R, Cuenot N. Seismicity induced during the development of the Rittershoffen geothermal field, France. Geothermal Energy. 2020;**8**(1):5

[16] Baujard C, Maurer V, Dalmais E, Ravier G, Glaas C, Genter A. Low induced seismicity in fractured geothermal reservoirs: Experience and lessons learned from Rittershoffen and Soultz-sous-Forêts plants operation and Illkirch reservoir development. In: Proceedings of European Geothermal Congress 2022. Berlin, Germany: European Geothermal Energy Council; 2022. p. 10

[17] Küperkoch L, Olbert K, Meier T. Long-term monitoring of induced seismicity at the Insheim geothermal site, Germany. Bulletin of the Seismological Society of America. 2018;**108**(6):3668-3683

[18] Lagache L, Genter A, Baumgaertner J, Cuenot N, Koelbel T, Texier P, et al. How is evaluated acceptability of an EGS project in Europe: The Soultz-Kutzenhausen geothermal project? In: Proceedings European Geothermal Congress. Pisa, Italy; 2013. pp. 1-4

[19] Chavot P, Heimlich C, Masseran A, Serrano Y, Zoungrana J, Bodin C. Social shaping of deep geothermal projects in Alsace: Politics, stakeholder attitudes and local democracy. Geothermal Energy. 2018;**6**(1):1-21

[20] Chavot P, Masseran A, Bodin C, Serrano Y, Zoungrana J. Geothermal energy in France. A resource fairly accepted for heating but controversial for high-energy power plants. In: Manzella A, Allansdottir A, Pellizzone A, editors. Geothermal Energy and Society. Springer International Publishing; 2019. pp. 105-122

[21] Cuenot N, Dorbath C, Dorbath L. Analysis of the microseismicity induced by fluid injections at the EGS site of Soultz-sous-Forêts (Alsace, France): Implications for the characterization of the geothermal reservoir properties. Pure and Applied Geophysics. 2008;**165**(5):797-828

[22] Charléty J, Cuenot N, Dorbath L, Dorbath C, Haessler H, Frogneux M. Large earthquakes during hydraulic stimulations at the geothermal site of Soultz-sous-Forêts. International Journal of Rock Mechanics and Mining Sciences. 2007;**44**(8):1091-1105

[23] Dorbath L, Cuenot N, Genter A, Frogneux M. Seismic response of the fractured and faulted granite of Soultz-sous-Forêts (France) to 5 km deep massive water injections. Geophysical Journal International. 2009;**177**(2):653-675

[24] Deichmann N, Giardini D. Earthquakes induced by the stimulation of an enhanced geothermal system below Basel (Switzerland). Seismological Research Letters. 2009;**80**(5):784-798

[25] Gaucher E, Maurer V. Évolution du suivi sismique des sites géothermiques en exploitation dans le Fossé rhénan supérieur. Géologues Géosciences et Société. 2021;**210**:94-102

[26] Wiemer S, Kraft T, Trutnevyte E, Roth P. Good Practice Guide for Managing Induced Seismicity in Deep Geothermal Energy Projects in Switzerland. Zurich: ETH Zurich; 2017

[27] Ravier G, Baujard C, Dalmais E, Maurer V, Cuenot N. Towards a comprehensive environmental monitoring of a geothermal power plant in the Rhine graben. In: Proceedings of European Geothermal Congress 2016. Strasbourg, France; 2016. p. 9

[28] Sanjuan B, Millot R, Innocent C, Dezayes C, Scheiber J, Brach M. Major geochemical characteristics of geothermal brines from the upper Rhine graben granitic basement with constraints on temperature and circulation. Chemical Geology. 2016;**428**:27-47

[29] Ledésert BA, Hébert RL, Mouchot J, Bosia C, Ravier G, Seibel O, et al. Scaling in a geothermal heat exchanger at Soultz-sous-Forêts (upper Rhine graben, France): A XRD and SEM-EDS characterization of sulfide precipitates. Geosciences. 2021;**11**(7):271

[30] Seibel O, Mouchot J, Ravier G, Ledésert B, Sengelen X, Hebert R, et al. Optimised valorisation of the geothermal resources for EGS plants in the upper Rhine graben. In: Proceedings World Geothermal Congress 2020+1. Reykjavik, Iceland; 2021. p. 9

[31] Eggeling L, Genter A, Kölbel T, Münch W. Impact of natural radionuclides on geothermal exploitation in the upper Rhine graben. Geothermics. 2013;**47**:80-88

[32] Scheiber J, Nitschke F, Seibt A, Genter A. Geochemical and mineralogical monitoring of the geothermal power plant in Soultz-sous-Forêts (France). In: Proceedings, Thirty-Seventh Workshop on Geothermal Reservoir Engineering. Stanford, California: Stanford University; 2012. pp. 1-10

[33] Mouchot J, Genter A, Cuenot N, Seibel O, Scheiber J, Bosia C, et al. First year of operation from EGS geothermal plants in Alsace, France: scaling issues. In: Proceedings, 43rd Workshop on Geothermal Reservoir Engineering. Stanford, California: Stanford University; 2018. p. 12

[34] Bosia C, Mouchot J, Ravier G, Seibt A, Jähnichen S, Degering D, et al. Evolution of brine geochemical composition during operation of EGS geothermal plants (Alsace, France). In: Stanford University. Stanford: California; 2021. pp. 1-21

[35] (2006) NF EN ISO 14044: Environmental Management—Life Cycle Assessment—Requirements and Guidelines

[36] (2006) NF EN ISO 14040: Environmental Management-Life Cycle Assessment-Principles and Framework

[37] Lacirignola M, Blanc I. Environmental analysis of practical design options for enhanced geothermal systems (EGS) through life-cycle assessment. Renewable Energy. 2013;**50**:901-914

[38] Lacirignola M, Meany BH, Padey P, Blanc I. A simplified model for the estimation of life-cycle greenhouse gas emissions of enhanced geothermal systems. Geothermal Energy. 2014;**2**(1):8

[39] Pratiwi A, Ravier G, Genter A. Life-cycle climate-change impact assessment of enhanced geothermal system plants in the upper Rhine Valley. Geothermics. 2018;**75**:26-39

[40] Parisi ML, Douziech M, Tosti L, Pérez-López P, Mendecka B, Ulgiati S,

et al. Definition of LCA guidelines in the geothermal sector to enhance result comparability. Energies. 2020;**13**(3534):1-18

[41] Douziech M, Tosti L, Ferrara N, Parisi ML, Pérez-López P, Ravier G. Applying harmonised geothermal life cycle assessment guidelines to the Rittershoffen geothermal heat plant. Energies. 2021;**14**(13):3820

[42] Douziech M, Ravier G, Ferrara N, Damen L, Sigurjonsson HA, Pérez-López P, et al. Simplified parameterized models for a multi-criteria environmental impact assessment of four types of geothermal installations. In: Proceedings World Geothermal Congress 2020+1. Reykjavik, Iceland: International Geothermal Association; 2021. p. 11

[43] Perez P, Ravier G, Pratiwi A, Genter A, Blanc I. Life Cycle Assessment and economic impacts of the Rittershoffen EGS geothermal plant, Upper Rhine Graben, France. In: Proceedings World Geothermal Congress 2020+1. Reykjavik, Iceland; 2021

[44] ADEME. Coût des énergies renouvelables en France. 2020

[45] Dalmais E, Genter A, Trullenque G, Leoutre E, Leiss B, Bär K, et al. MEET project: Toward large scale deployment of deep geothermal energy in Europe. In: Proceedings World Geothermal Congress 2020+1. Reykjavik, Iceland; 2021. pp. 1-11

[46] Kunan P, Ravier G, Dalmais E, Ducousso M, Cezac P. Thermodynamic and kinetic modelling of scales formation at the Soultz-sous-Forêts geothermal power plant. Geosciences. 2021;**11**(12):483

[47] Abbas A, Seibel O, Ravier G, Leoutre E. Roadmap for heat production from EU main geological contexts. Open File Report, MEET Project, Deliverable. 2022;**D6**:15

[48] Gallup DL. Geochemistry of geothermal fluids and well scales, and potential for mineral recovery. Ore Geology Reviews. 1998;**12**(4):225-236

[49] Bourcier WL, Lin M, Nix G. Recovery of minerals and metals from geothermal fluids. In: Proceedings, SME Annual Meeting. Cincinnati, OH, United States: Society for Mining, Metallurgy and Exploration; 2003

[50] European Commission, Directorate-General for Internal Market, Industry, Entrepreneurship and SMEs. Communication from the Commission to the European Parliament, the Council, the European Economic and Social Committee of the Regions. Critical Raw Materials Resilience: Charting a Path towards greater Security and Sustainability; 2020

[51] Bloomquist RG. Economic benefits of mineral extraction from geothermal brines. In: Proceedings Sohn International Symposium; Advanced Processing of Metals and Materials. Vol. 6. 2006. pp. 553-558

[52] Harrison S. Technologies for extracting valuable metals and compounds from geothermal fluids. In: Final Report for Department of Energy Geothermal Technologies Program Grant DE--EE0002790, United States. 2014

[53] Alessia A, Alessandro B, Maria V-G, Carlos V-A, Francesca B. Challenges for sustainable lithium supply: A critical review. Journal of Cleaner Production. 2021;**300**:126954

[54] Bale M, May A. Processing of ores to produce tantalum and lithium. Minerals Engineering. 1989;**2**(3):299-320

[55] Sanjuan B, Gourcerol B, Millot R, Rettenmaier D, Jeandel E, Rombaut A. Lithium-rich geothermal brines in Europe: An up-date about geochemical characteristics and implications for potential Li resources. Geothermics. 2022;**101**:102385

[56] Gloaguen E, Melleton J, Lefebvre G, Tourière B, Yart S, Gourcerol B. Ressources métropolitaines en lithium et analyse du potentiel par méthodes de prédictivité. Open File Report. 2018. BRGM/RP-68321-FR

[57] Fries D, Lebouil S, Maurer V, Martin C, Baujard C, Ravier G, et al. Lithium extraction through pilot scale tests under real geothermal conditions of the Upper Rhine Graben. In: Proceedings European Geothermal Congress 2022. Berlin, Germany: European Geothermal Energy Council; 2022. p. 7

[58] Rettenmaier D, Zorn R, van den Kamp J. Business and sensitivity analysis. Eugeli Project report. 2021

Chapter 5

General Information on Geothermal Energy

Héctor Aviña-Jiménez, Eduardo Pérez-González and Rodrigo Alarcón-Flores

Abstract

The use of natural resources in a more responsible and comprehensive manner has become more relevant in recent years. The energy crisis and climate change have targeted the development of technologies that allow the use of renewable energies with greater performance, efficiency, and results. Geothermal energy plays an important role since it is available 24 hours a day, 365 days a year however, it represents a great challenge since its extraction is not a trivial fact, that is, every day it is necessary to further improve exploration techniques and exploitation to access increasingly deeper resources and greater energy potential. This section addresses the different applications that have been developed throughout the world and that serve as parameters and guides for their replication in Latin America, since due to its geothermal potential it has various opportunities to develop technology and agro-industrial production processes that are of Vital importance for human development in the coming decades.

Keywords: low enthalpy, direct uses, Cascade and cogeneration uses, heat pumps, binary cycles

1. Introduction

Geothermal energy is used worldwide for the generation of electricity directly from geothermal heat through the so-called direct uses; if it is used properly, it is possible to obtain an integral use of the Geothermal and its resource. Mexico and Latin America have enormous potential for Geothermal resources; however, successful projects for the comprehensive use of geothermal resources, such as the geothermal food dehydrator in Mexico, are just beginning to materialize.

Currently, the use of geothermal energy applied to direct uses has been an alternative to be used in industry, in general in other parts of the world, obtaining multiple benefits, and making geothermal projects sustainable [1–6].

The opportunities for exploiting medium and low-temperature geothermal resources (<180°C) have great potential in Mexico and Latin America. These opportunities range from low-tech balneology to air conditioning applications and industrial services. In 1959, Mexico was a pioneer in Latin America in the exploitation of geothermal energy, however, it has lagged significantly in terms of the comprehensive use of direct uses of geothermal energy. For this reason, the iiDEA Group of the Institute of Engineering of the National Autonomous University of Mexico (UNAM) developed a methodology to achieve projects for the geothermal energy. Including several areas of specialization to obtain a successful and sustainable project, giving greater importance to the social components of the projects, and directing them to the benefit of local communities. This article describes the method to integrate technical, environmental, and legal frameworks with economic, social, and political characteristics. This method offers a guide to mitigate the barriers and challenges that prevent the development of direct use of geothermal projects, according to the specific needs of each geothermal resource and the local community that surrounds it. For example, technical issues such as temperature and mass flow of the resource; commercial issues such as the product's marketing channel; political issues such as government support and the lack of specialized technical advice; and most importantly, include the local community in the operation and business of the projects.

2. Principles of geothermal energy

The thermal energy coming from the interior of the Earth is manifested indirectly through volcanism, thermal gradients in the ground, displacement of tectonic plates, and superficial geothermal emanations: lava, boiling mud pools, fumaroles, geysers, and thermal waters. Geothermal energy, as indicated by its etymological origin (lati. Geo-earth and thermo-heat), has already been explained and can be defined as the heat stored inside the earth. Its origin is associated with four sources:

1. Proto-Earth: Stage of formation of the planet Earth 6400 million years ago. Thermal energy is generated by shocks or impacts associated with meteorites that shaped the planet Earth. Heavy and light elements separate to form the core, mantle, and crust. The heat has been preserved by the insulating effect of the rock, gradually manifesting toward the surface.

2. Radiogenesis: Radioactive elements such as uranium, thorium, rubidium, and potassium generate 60% of the heat in the crust, where they are present in abundance compared to the mantle because their large atomic radii are less compatible with structures, minerals of said terrestrial layer; however, it is estimated that about 47 MWt of the mantle are generated by radiogenesis [7].

3. Gravitational pressure: Similar to Charles's Law, where the temperature of a gas is increased by its compression, something similar happens in solids, except that the change in volume is not as evident as in gases. The heat is generated by the internal gravitational attraction or compression of the rocks, accumulating the heat in the depths.

4. Earthquake fault friction: Manifestation of local frictional heat along earthquake faults; the heat generated is enough to partially melt the rock (pseudotachylite). Most geothermal power plants harness the energy produced at these active faults [7].

2.1 Geothermal systems

They are different geothermal systems, of which hydrothermal systems are of great relevance because they are commercially exploited through geothermal fields with particular characteristics between each of them. The rest of the systems found in the earth's crust, which are expected to be exploited once the technology are perfected, are geopressurized systems, magma chambers, and hot dry rock. The main characteristics of each of them are presented below.

2.1.1 Hydrothermal systems

They are very unique cases, considered as rare hydrothermal places; they represent less than 10% of the geothermal systems around the world. They are characterized by the accumulation of water, steam, or both, and this is what allows the conduction of heat from the most deep settlement toward the surface. Such a system has permeable formations that contain the fluid and is located within the economic range of drilling platforms so that some of the fluids can be brought to the surface for useful applications. The geothermal fields, through which the resource is exploited, can be:

i. Semi-thermal: Produces hot water up to 100°C at a depth of 1 or 2 km

ii. Hyperthermals: They can be wet or dry, and are sometimes called water-dominant and steam-dominant fields, respectively.

- Wet: They produce water under pressure at more than 100°C, while its pressure is lowered and removed, a fraction evaporates and turns into steam.

- Dry: They produce dry saturated steam or slightly superheated at pressures above atmospheric.

2.1.2 Geopressurized

These fields are filled with pressurized hot water, its pressure exceeds 40 to 90% of the hydrostatic pressure corresponding to its depth. They are found at great depths (up to about 6000 m in some places), so drilling costs are high.

Magma chambers: Located at different depths, there are pockets of magma with temperatures between 600 and 1300°C, which is why they are considered manifestations with high energy content. However, they are not the usual ones and the technology allows them to handle high temperatures.

2.1.3 Hot dry rock at moderate depths

About 50 to 60% of the Earth has these manifestations of heat (between deep or superficial), from can already be extracted through methods that have been successfully tested, but the process still needs to be perfected to it to the level a commercial. The first research works were carried out in the 1970s and were called enhanced geothermal systems (EGSs). It consist of creating an artificial reservoir of hot water for its extraction in the form of steam. The Soultz project in France was a pioneer in the development of EGS and of which the technical feasibility was demonstrated; however, it was only possible to produce 50% of the expected steam.

2.1.4 Hot dry rock at great depths

It is a resource with greater abundance than the previous ones. Due to the great depth at which they are located, they are currently economically unfeasible. Surely derived from the improvement of the EGS systems, the first studies for their use can be developed.

2.1.5 Hydrothermal vents

There are submarine hydrothermal systems, with temperatures of up to 300°C, they can be shallow manifestations (depth < 200m) or deep; examples of surface manifestations in the world are found in Planty Bay, New Zealand; Ambitle and Lithir hydrothermal system, Papua, New Guinea; Cala Karaternaya, Russia; Milos Island, Aegean Sea; Eolian Islands, Italy; Punta Banda to the west of the Baja California Peninsula, the western part of the Gulf of California in Bahía Concepción and the central part of the Pacific coast of Mexico. Its main characteristics are the discharges that apparently have a considerable effect on the adjacent biological communities (benthic, planktonic, and phytoplanktonic), presenting a good adaptation to this type of habitat, with high concentrations of nutrients, heavy metals, and traces. They form along underwater fractures, so they align with them, forming groups in a row with separations of up to 20 m between each one. Currently there is no technology that allows its commercial use, and the way in which it would affect the biotic system that depends on these underwater manifestations has not been tested. Several prototypes have been designed and tested to generate electricity, developing organic Rankine cycle (ORC), Seebeck effect electricity generation systems, as well as water desalination processes, mineral extraction, and biofuel generation.

3. Installed potential and current energy consumption in the world

Variables in the properties of matter at high depths and pressures are not easily replicable in laboratories, so they are subject to estimation. H. Cristopher and H. Armstead (1989), cited in Ref. [7], estimated usable energy of 79 PJ. To arrive at this estimate, the approximate averages of the specific heat and thermal conductivity of the material under these conditions, the local differences in the thickness of the crust and the density of the rock were considered. It is worth mentioning that in this calculation only the energy contained in the crust was considered, so all the chemical energy present in the form of fuel, all the energy emitted by radioactive rocks, and all the heat conducted from the hot mantle were discarded.

The total heat of the crust could satisfy a large part of man's energy needs for a long time, coming to be considered, in a certain sense, so abundant as to be a practically infinite source of energy; however, on a local scale, a field can be depleted by prolonged exploitation. It has been documented that for a field to maintain its production, it depends on drilling more wells in an area that is always expanding, reaching the limits of the geothermal system and then gradually decreasing its production. Therefore, it is important to continue with the development of technology that allows free access to 90% of the geothermal energy of the crust that is currently not exploited. EGS for

Country	Project	Developer/consortium	I + D	Pilot and demo
Electricity Generation				
Korea	Pohang	DESTRESS		✓
Germany	Groß Schönebeck	DESTRESS		✓
Swiss	Haute-Sorne	DESTRESS	—	—
	Bedretto	DESTRESS	✓	
	Etzwilen	Geo-Energie Suisse AG	—	—
	Triengen	Geo-Energie Suisse AG	—	—
	Pfaffnau	Geo-Energie Suisse AG	—	—
	Avanches	Geo-Energie Suisse AG	—	—
France	Soultz-sous-Forêts	DESTRESS		✓
Australia	Embalse Habanero	Geodynamics Ltd.		✓
	Depósito Parlana	Petratherm Ltd.		✓
USA	FORGE	Geothermal Technologies Office of the USDOE	✓	
European Comission	GEOFIT	Consortium	✓	
	MEET	Consortium		✓
Mexico	GEMex	SENER - CONACYT - UE	✓	
Spain	Proyecto Güimar	Arllen Development** y ayuntamiento Güimar	✓	
District Heating/Greenhouses				
Lithuania	Klaipėda	DESTRESS		✓
Hungary	Mezőberény	DESTRESS		✓
Netherlands	Westland	DESTRESS		✓
	Middenmeer	DESTRESS		✓
Iceland	Geldinganes	DESTRESS	✓	
France	Rittershoffen	DESTRESS	✓	

*— Canceled. **Slovak multinational.*

Table 1.
EGS projects in the world.

heat extraction from hot and dry rock systems is being researched and developed in various parts of the world, the most representatives are presented in **Table 1**.

USDOE: Department of Energy EE. UU.; SENER: Secretary of Energy, Mexico; CONACYT: National Council for Science and Technology, Mexico; EU: European Union; DESTRESS: Demonstration of soft stimulation treatments of geothermal reservoirs. (Consortium).

The most exploited geothermal systems are hydrothermal and, to a lesser extent, EGS. The electrical generation cycles are by condensation, back pressure, ORC cycles, and kalina. The installed capacity in the world for the generation of electricity grows an average of 11% per year.

However, the production of electrical energy was not the first activity that was developed for the use of the heat of the earth; the first application was balneology, later the cooking of food and therapeutic applications. There is currently a record of 997 GWt installed for different applications, all of them cataloged under the generic name of direct uses.

4. Use of geothermal energy

Direct uses (DUs) are applications for thermal use in residential, commercial, and industrial sectors; Where before the heat from the burning of L.P. gas, natural gas, firewood, coal, or other was used, it is now replaced by the heat of the Earth. However, there is the development of electricity generation systems with moderate temperatures, up to 200°C, for local consumption and/or applied to the aforementioned sectors. Geothermal energy is classified into three groups, high ($t > 200°C$), medium ($90 \leq t < 200°C$), and low enthalpy ($30 \leq t < 90°C$). It is important to mention that the medium and low enthalpy scales can be relative, and the above depends on the environmental temperatures of each region; for very cold countries with temperatures below zero, having water at 10 °C from the Earth can be considered geothermal, but in very hot regions with temperatures up to 45°C, it may not be classified as such.

Based on this definition, the use of the resource for its various applications is regulated and controlled. Currently, the Lindal diagram is well known in honor of Baldur Lindal [8], who documented a series of geothermal applications based on the amount of energy/temperature/enthalpy of the resource, but over time it has been updated along with the evolution of industrial activities.

In search of more efficient use of medium and low enthalpy energy, a series of DUs are included that operate sequentially in the order of the energy level they require. The first applications are those that need more temperature, then those of intermediate temperature, and finally those of low temperature, until all the available thermal

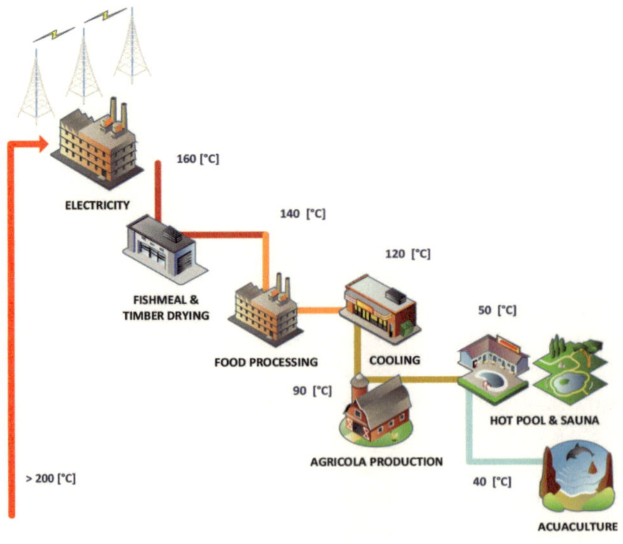

Figure 1.
Cascading usage concept.

energy is extracted, with ambient temperature being the lower limit; quality and direction of energy are concepts explained by the second law of thermodynamics and that are perfectly exemplified through this practice called cascading uses (CU) (**Figure 1**).

In 1995, only 28 countries used geothermal energy; in 2000, 58; 2005, 72; 2010, 78, 2015, 82, and finally in 2020, the figure reported was 87, with the participation of Bolivia, Burundi, Cyprus, Faroe Islands, Malawi, Malaysia, and Nigeria. With this growth in the use of geothermal energy (see **Table 2**), energy savings are 24.4 million tons of oil equivalent (TOE) per year (167 million barrels), leaving 36 million tons burning coal (96

	Country	1995 MWe	2000 MWe	2005 MWe	2010 MWe	2015 MWe	2020 MWe
	Total	**6866.80**	**7974.10**	**9067.10**	**10,716.70**	**12,635.10**	**15,847.2**
1	USA	2816.7	2228.0	2544.0	3093.0	3450.0	3700.0
2	Indonesia	309.8	589.5	797.0	1197.0	1340.0	2289.0
3	Philippines	1227.0	1909.0	1931.0	1904.0	1870.0	1918.0
4	Turkey	20.4	20.4	20.4	82.0	397.0	1549.0
5	New Zealand	286.0	437.0	435.0	628.0	1005.0	1064.0
6	Mexico	753.0	755.0	956.0	958.0	1017.0	1005.8
7	Italy	631.7	785.0	790.0	843.0	916.0	916.0
8	Iceland	50.0	170.0	322.0	575.0	665.0	755.0
9	Kenya	45.0	45.0	127.0	167.0	594.0	1193.0
10	Japan	413.7	546.9	535.0	536.0	519.0	550.0
11	Costa Rica	55.0	142.5	163.0	166.0	207.0	262.0
12	El Salvador	105.0	161.0	151.0	204.0	204.0	204.0
13	Nicaragua	70.0	70.0	77.0	88.0	159.0	159.0
14	Russia	11.0	23.0	79.0	82.0	82.0	82.0
15	Papua New Guinea	0.0	0.0	39.0	56.0	50.0	11.0
16	Guatemala	33.4	33.4	33.0	52.0	52.0	52.0
17	Germany	0.0	0.0	0.2	6.6	27.0	43.0
18	Portugal*	5.0	16.0	16.0	29.0	28.0	33.0
19	China	28.8	29.2	28.0	24.0	27.0	34.9.0
20	France**	4.2	4.2	15.0	16.0	16.0	17.0
21	Ethiopia	0.0	8.5	7.0	7.3	7.3	7.3
22	Austria	0.0	0.0	1.0	1.4	1.2	1.3
23	Australia	0.2	0.2	0.2	1.1	1.1	0.6
24	Argentina	0.6	0.0	0.0	0.0	0.0	0.0
25	Romania	0.0	0.0	0.0	0.0	0.1	0.0
26	Taiwan	0.0	0.0	0.0	0.0	0.1	0.3
27	Thailand	0.3	0.3	0.3	0.3	0.3	0.0

Table 2.
IGA data [9, 10].

CODE	Capacity MWt					
	2020	**2015**	**2010**	**2005**	**2000**	**1995**
GHP	82,319.61	50,258.00	33,134.00	15,384.00	5275.00	1854.00
H, & D	6113.02	7602.00	5394.00	4366.00	3263.00	2579.00
G	9124.37	1972.00	1544.00	1404.00	1246.00	1085.00
F	949.67	696.00	653.00	616.00	605.00	1097.00
A	256.66	161.00	125.00	157.00	74.00	67.00
I	852.77	614.00	533.00	484.00	474.00	544.00
B	24,190.42	9143.00	6700.00	5401.00	3957.00	1085.00
C & S	435.07	360.00	368.00	371.00	114.00	115.00
O & K	110.99	79.00	42.00	86.00	137.00	238.00
Total	**124,352.58**	**70,885.00**	**48,493.00**	**28,269.00**	**15,145.00**	**8664.00**

I = Industrial process heat; H = Individual space heating (other than heat pumps); C = Air conditioning (cooling); D = District heating (other than heat pumps); A = Agricultural drying (grain, fruit, vegetables); B = Bathing and swimming (including balneology); F = Fish farming; G = Greenhouse and soil heating; K = Animal farming; O = Other (please specify by footnote); S = Snow melting; GHP = Geothermal Heat Pumps.

Table 3.
Installed capacity in TJ/year per application developed in the world [3, 11].

million tons of CO_2). Worldwide, there are 279 cases of DU, and some of them integrate an entire CU system; the distribution of said DU can be seen in **Table 3**.

4.1 Direct uses, background, and current situation

Despite the fear generated in the world by the most violent manifestations of geothermal energy, such as volcanoes, due to the destruction of Pompeii and Herculaneum [7], it did not take long for humanity to explore the benefits of thermal waters, since the appearance of the belief in drinking water as a prophylactic remedy for its healing and laxative properties, spreading and popularizing these practices, thus giving rise to balneology.

4.1.1 Balneology

It flourished at the time of the Roman Empire; however, it was a practice that came from the Greeks. Baths, as they were known, were established as meeting places, in some ways comparable to the coffee houses of the eighteenth century London. In North America, Paleo-Indians used minerals for medicinal purposes and hot springs were neutral zones where members of warring nations should bathe together in peace.

Spas, hydro-treatments, and jet baths spread throughout Europe and elsewhere, frequented by invalids, hypochondriacs, and those who simply flocked to them for pleasure. Subsequently, the concept of SPA emerged, which comes from the Latin abbreviation S: *sanitas*, P: *per*, A: *aquas* [12], or health through water. In the nineteenth century, SPAs in Europe were very sophisticated and elegant centers, like the famous spas in France, Germany, and Burma.

During World War II, in Rotorua, New Zealand, the Queen Elizabeth Hospital used mineral waters and mud from hot springs to help soldiers in the Pacific wars recover from war wounds. Currently, in European countries and Japan, they have specialist doctors who attend spas, with the aim of treating or preventing different diseases [13].

Yasuhiro Ishikawa, better known as Dr. Bath, in Japan, states that the typical bath is necessary to maintain proper hygiene, but it is not enough to improve the immune system, prevent diseases and maintain physical and mental health as is achieved with baths, hot springs baths enriched in salts. The effect of these salts has been studied, and the adequate retention of heat has been verified by means of a thermographic camera, activating the cells that increase heat-shock proteins that delay the production of milk acid in the muscles.

Currently, of the 87 countries that take advantage of low enthalpy geothermal energy, 72 do so through authorized thermal centers, mainly for tourism. Between all of them, there is an installed potential of 24,190 MWt.

4.1.2 Mineral extraction

After the development of balneology and some culinary practices such as cooking fish, eggs, and some vegetables in geothermal fumaroles, the extraction of minerals and salts present in the hot water was developed. In 1818, in Larderello, Tuscany (Italy), boron salts were exploited for the first time for industrial purposes; the evaporation was done inside a brick dome, which was known as Covered Lagoon. Near the dome, deep wells were drilled to access hot water with high concentrations of boron. In 1827, the founder of this industry, the French Francois Larderel, developed a system to take advantage of the heat of the fluids in the evaporation process, leaving aside the combustion of wood. It was considered the most important industry in Europe, and by then sulfur, vitriol, alum, and boric acid were already being extracted. Over time, however, electricity generation became more valuable than geothermal mining.

Currently, a new industry is growing driven by the new energy economy, and it is due to the great demand for lithium that has been increasing at an unprecedented rate. For the next decade alone, an increase in demand of just under 10 times the current one is expected (in 2018, the demand was 150 thousand tons, by 2028 an increase to 1.5 million tons is expected), a demand that cannot be satisfied with conventional lithium extraction processes, since they are expensive and cannot generate the production volumes necessary for a future in which the vehicle fleet will be mostly electric.

Unlike current extraction methods, geothermal lithium extraction is a closed-loop system that returns spent brine to its original source; and by virtue of the fact that it is an activity derived from electricity generation, it benefits from the electrical energy produced in these plants, so it is a process that operates with 100% renewable energy. In a matter of hours, not months, compared to traditional technology, it produces high-purity lithium; its environmental advantages are its small carbon footprint, close to zero, it is not dependent on weather or water, it does not require open pit mines or large evaporation ponds, and it operates 24/7.

Currently, lithium production is concentrated in Australia, China, and South America (Argentina, Chile, and Bolivia). According to Benchmark Mineral Intelligence, the United States currently produces 1% of the world's raw lithium materials and only 7% of its lithium chemicals, while China, Japan, and South Korea produce 85% of the chemicals of lithium needed to power electric vehicles. For this reason, the US has recognized the need to develop a strategy around its critical mineral security.

4.1.3 Domestic service: Heating and hot water supply at the district level

The Greeks and, later, the Romans were the ones who left examples of applications with geothermal energy, as already mentioned, for which hot water was distributed through an open network that connected to the basements of buildings. These practices were spread by the Romans and eventually reached Japan, America, and Europe.

In 1332, in Chaudes-Aigües, France, a new distribution system with hollow logs was installed, serving 30 houses, as well as activities related to the washing of wool and fur. Also in France, but in 1833, in the Grenelle district, Paris, the first borehole began, through an artesian well 548 m deep, which took 8 years to build and from which drinking water at 30°C was extracted.

The first modern district heating network powered by geothermal energy was installed in Reykjavík, Iceland, in 1930. Since then, this innovative heating system has spread throughout the world, with plants installed in France, Italy, Hungary, Romania, Russia, Turkey, Georgia, China, the United States, and Iceland, in Iceland itself, where currently 95% of its inhabitants have heating through a network of 700 km of insulated pipes that transport hot water. In 1947, Kemler, E.N., in his publication "Methods of earth Heat Recovery for the Heat Pump" already showed the diagrams of the different connection methods of heat pumps.

After the Second World War and due to the expansion of other cheaper energy sources, mainly petroleum derivatives, this system was left aside. District heating networks continued to be installed in Europe during this period, mainly in the Nordic countries due to the shortage of natural gas and electricity. In the 1970s, with the oil crisis, district heating networks regained their importance, especially in the United States, as well as in Northern Europe, Russia, Japan, China, and Korea, initiating an intense activity of exploration and investigation of geothermal resources in order to use them for the production of electrical energy or for heating and hot water. In this way, the development of geothermal energy was stimulated, and global geothermal production increased from 400 Wt in 1960 to 15,847.2 MWt in 2020.

These heating systems, widely used in Europe, represent 30% of the total energy used for conditioning spaces and water, benefiting 75% of all buildings that have heating [14]. The main countries with high installed capacity for this geothermal use are listed in **Table 4**.

On the other hand, the heating of spaces through geothermal heat pumps (GHP) is presented independently because they normally work with a fluid that is not necessarily geothermal; however, because the heat transfer is carried out with a constant temperature from the earth, somehow it is still considered as energy coming from the Earth. A GHP consists of a closed system of high-density polyethylene pipes through which water (not geothermal water) flows; This fluid transfers or absorbs energy from the earth depending on the season of the year. The leading countries in the implementation of this type of system are listed in **Table 5**.

4.1.4 Aquaculture and agricultural products

Aquaculture is defined as the farming or rearing of aquatic species, such as catfish, tilapia, sturgeon, largemouth bass, shrimp, tropical fish, crustaceans, molluscs, aquatic plants, and even alligators. This activity is carried out under controlled conditions with the aim of favoring the development of the specimens. The use of geothermal energy directly or indirectly in heating the habitat of the species depends on the quality of the water since normally the geothermal water or brine has dissolved salts

No	Country	Capacity (MWt)
1	Iceland	1650.00
2	Turkey	1453.00
3	France	510.00
4	Germany	349.54
5	China	346.00
6	Hungary	300.60
7	Italy	225.00
8	Russia	220.00
9	Japan	203.34
10	USA	179.03
	Others (30 countries)	676.52

Distinto a GHP.

Table 4.
Top 10 countries with the largest installed capacity for heating, 2020.

No	Country	Capacity (MWt)
1	China	26,450.00
2	USA	20,230.00
3	Sweden	6680.00
4	Albania	4497.00
5	Germany	4400.00
6	Finland	2300.00
7	Switzerland	2172.00
8	France	2015.00
9	Canada	1822.50
10	Ukraine	1600.00
	Others (44 countries)	10,153.11

Table 5.
Top 10 countries with installed capacity in GHP.

and minerals that can be harmful. A heat exchanger is generally used to transmit this energy to the water in the ponds. The rearing of species in controlled warm environments affects chemical and biological processes (such as metabolism) favoring larger and more developed specimens in less time, an ideal practice in ectothermic[1] organisms. **Table 6** shows the top 10 countries with aquaculture development.

Regarding the development of agriculture through heated greenhouses with geothermal energy, either directly or through the use of GHP, it represents a great

[1] Ectothermy is the condition of a group of living beings that are not capable of generating, through various metabolic or physiological processes, their own internal heat. In this way, they must depend on external heat sources to reach a certain body temperature, reducing their activity when the environmental temperature is not adequate.

No	Country (MWt)	Ref.	Project
1	China (420.00)	[15, 16]	There are 14 units that use geothermal water to raise left mouth fish, grouper, tilapia, white-backed turtles, crabs, and prawns. The reported aquaculture area is 6.47 × 105 m^2. A total of 300 aquaculture farms are reported and represent 4.45 × 106 m^2 throughout the country. It saw strong growth in Beijing, Tianjin, Fujian, and Guangdong, as well as 20 other provinces, mainly Guangdong and Fujian. Of the total geothermal water used in the country, only 5.7% is used for aquaculture.
2	Italy (130.00)	[17, 18]	An installed capacity of 1425 MWt for direct uses is reported, of which 4.6% corresponds to aquaculture. The projects are located in Orbetello (Tuscan coast) and the Apulian coast. It is considered the most important aquaculture industry in Europe. The production of Orbetello alone is estimated at 900 tons/year with a staff of 60 workers.
3	USA (122.13)	[19–24]	Liskey Farms Inc. It is a company dedicated to the breeding of tropical fish. The temperature of the 37 ponds is 23°C (74°F). Idaho Fish Breeders Inc. It is dedicated to the breeding of catfish for its commercialization. Their activity began in the early 1970s. They maintain a constant temperature in the ponds between 27 and 29°C. Oregon Institute of Technology. Research began in 1975 with the idea of harnessing the remaining thermal energy from the University's heating system. Giant freshwater prawns (*Macrobrachium rosenbergii*) are farmed. The goal is to have a constant temperature in the ponds around 27 ± 1°C. Fort Indian Community, California. They market catfish, mainly in San Francisco. They use geothermal water at 40°C, mix it with fresh water, and heat the ponds to 27°C. Various projects. Throughout the country, 27 projects are reported with an installed capacity of 122.13 MWt.
4	Iceland (110.00)	[25–27]	In 2013, there were 70 aquaculture parks that together produced 70 thousand tons of salmon, mainly arctic char and trout. Of those plants, only 20 were powered by geothermal energy. Currently, the main producers are two. The first is the Íslandslax plant, belonging to Samherji company. It has 2000 m^2 and a water volume of 1500 m^3. For its process, 10 kg/s of geothermal water at 90°C is required. The other plant is Silfurstjarnan located in the north of the country in Öxarfjördur. It annually produces thousand of tons.
5	Israel (31.40)	[28, 29]	Aquaculture in Israel is a unique practice in that it has been developed throughout the Negev desert; there are about 15 aquaculture centers, established since 1980. These aquaculture centers produce fish for human consumption (barramundi (*Lates calcarifer*), red croaker (*Sciaenops ocellatu*), European sea bass (*Dicentrarchus labrax*), North African catfish (*Clarias gariepinus*), Nile tilapia (*Orechromis niloticus*), ornamental fish and some crustaceans. Only ornamental fish are exported. Aquaculture is developed in two regions: 1. Jordan Valley at Hammat Gader Springs, where four springs of different temperatures provide warm water (27°C) for fish and shrimp farming. 2. It is found along the Mediterranean coast, where numerous ponds have been established. The hot water (26°C) comes from different shallow geothermal wells (30 m deep).

No	Country (MWt)	Ref.	Project
6	New Zealand (17.00)	[30]	The first aquaculture farm was established in Taupo, Wairakei in 1987. Fresh water from the Waikato River is conditioned with geothermal hot water at a constant temperature of 28°C. The study began with 20 males and five females imported from Malaysia; a year later, 25 males and 30 thousand post larvae were imported again from the same region.
7	Argelia (15.00)	[31–34]	Meanwhile, a production of 1700 tons/year of tilapia is estimated. The geothermal water used (from the Sahara desert) has a temperature of 40°C. The projects are 80% financed by the government as an initiative of the National Aquaculture Development Plan.
8	France (9.40)	[35]	Take advantage of the remaining energy from the heating of greenhouses (5 hectares) for the cultivation of eels and sturgeons. From a flow of 200 m^3/h of hot water at 75°C, cold water from the L'Eyre river is heated to condition it up to 17°C. Annual production of 10 tons of eels and 20 tons of sturgeons is estimated.
9	Japon (7.61)	[36]	Tilapia farming in a cold area of eastern Hokkaido, shrimp in Fukushima, and tiger puffer fish in Tochigi.
10	Argentina (7.03)	[37]	It has two aquaculture projects, one in the Greater region and another in Bahía Blanca.

Table 6.
Top 10 countries with aquaculture development.

opportunity for countries that do not have the right climatic conditions to produce certain foods throughout the year, or simply it would be impossible. An example of this, which at the time caused euphoria among Icelandic farmers, was in 1930, when the cultivation of tropical fruits, such as bananas, Began; Iceland currently has more than 200 thousand m^2 dedicated to greenhouses that supply fruits and vegetables (mushrooms, tomatoes, strawberries, flowers, and bananas) to the country's supermarkets, even allowing some export sales [38].

This practice has allowed the saving of 77 million m^3 of gas in the Netherlands, where the dependence of horticulture (greenhouses) on fossil fuels denotes the energy risk that can be solved with this type of energy [39].

The dehydration of food is another novelty, which consists of eliminating the moisture that food has. This can be done by different methods, and one of them is by hot air. This process consists of extracting heat from geothermal water through the use of a radiator (water–air heat exchanger), in this way the air is heated, which is then used to dry food, ranging from cereals, tubers, crops oilseeds, vegetables, spices, cocoa (*Theobroma cacao L.*) and coffee (*Coffea L. Rubiaceae*), fruits, and medicinal plants.

4.1.5 Industrial

The activities can be as varied and even as particular as enhanced oil recovery, mineral or rare earth leaching, metal mining (such as gold), mushroom cultivation, pulping for paper, drying of wood, and the tanning of wool or leather. Currently, the developments in this line and the new trend toward cogeneration processes or simply the creation of new production processes in order to optimize energy, agricultural, and natural resources in general, projects of singular interest have been developed (**Table 7**).

Country	Ref.	Project
Italy	[40–42]	Cheese factories, nurseries, and craft beer. Vapori di birra is a brewery that harnesses geothermal energy. Its two characteristic beers are Geyser and Solfurea e la Magma, each with its own characteristics and peculiarities.
Iceland	[43]	The hot water that flows into Urriðavatn is the only certified drinking hot water in Iceland, so it has been used to produce a special beer as well as geothermal tea.
Kenya	[44, 45]	The development of an industrial park is sought in Mai Mahiu, Naivasha, where government capital projects offer alternatives in new jobs and with productive activities that represent important income in the state. The GDC company is testing prototypes for: 1. milk pasteurization 2. greenhouses 3. aquaculture 4. laundries 5. cereal dryer
New Zealand	[46–51]	1. Imanaka (Japanese company) in association with a group of Maori entities has created the company Waiu Dairy (67% of the shares belong to 11 Maori entities and 33% to the Japanese company). Its objective is to produce 8 thousand metric tons of powdered milk with geothermal steam (4 to 6 tons/h at 16 Barg). 2. Oji Fiber Solutions (OjiFS) is a pulp mill in Tasman, Kawerau. It consumes from 20 to 50 tons/h of geothermal steam at 12 Barg. With this project, greenhouse gas emissions have been reduced by 10,000 tons of CO_2/year.
EUA	[52]	Klamath Basin Brewing Company opened in 2005 after renovating the historic Crater Lake Creamery building is located in Klamath Falls, Oregon. Currently, 10 different beers are brewed.

Table 7.
Examples of industrial development with geothermal energy.

4.1.6 Electricity generation with low-temperature geothermal resources

4.1.6.1 Historical panorama of electricity generation with geothermal energy

With the development of populations toward a modern civilization, geothermal energy was used to meet the energy needs that said civilization had been demanding. At the beginning of the twentieth century, Prince Piero Ginori Conti experimented with geothermal steam to produce electrical energy. After several years of experimentation, in 1904, he managed to turn on five light bulbs; he used a piston coupled to a 10 kWe dynamo. The system was powered by steam that was produced in a heat exchanger, which was fed with steam from the geothermal field near Larderello. The temperature of the resource was generally 150°C, with which pure water evaporated. Currently, it is possible to take advantage of geothermal resources with temperatures from 60 to 150°C using organic Rankine cycles, so the Principe Piero plant can be classified as the first binary cycle, water–water.

In 1908, a 20 kWe generating plant was installed, supplying power to the main industrial and residential buildings in Larderello. Five years later, the first commercial plant was built in Larderello, consisting of a 250 kWe turbine, manufactured by the Electromecánica Tosi company, which was designed to work with dry steam at

3 bar pressure at the wellhead. The steam was generated in a heat exchanger fed with geothermal steam at a temperature of 200 to 250°C. In 1923, the installed capacity in electricity generation was equalized with hydroelectric power plants; it was achieved with the generation of 3.5 MWe, and generated with the first direct geothermal power plant, that is, the steam produced was entered directly into the turbine, improving the use of the energy. Energy, without having to evaporate pure water. In 1930, there was an installed capacity of 12.15 MWe, of which 7.25 were generated in indirect cycles and 4.9 came from direct cycles.

The Larderello region was a strategic region, it provided electricity to the entire railway network of central Italy, for which it was bombed in the spring of 1944; together with all the geothermal power plants, chemical plants in the area and the production wells, with the exception of the 23 kWe well, which has served as a school well for the training of technical personnel from 1925 to date. After World War II, installed capacity recovered and by 1950 there were about 300 MWe. Until now, the technology developed only served to generate electrical energy with the dry steam that was produced, but in fields such as those in New Zealand, wet steam was available. In November 1958, the first groups of turbines were installed, five high pressure and two intermediates pressures, which used the wet steam that characterized the Wairakei geothermal field. The biphasic mixture was led to a cyclone separator, the resulting water was evaporated with pressure reduction, evaporating between 15 and 20%.

Over time, the steam that fed the high pressure turbines decreased, and currently only the intermediate pressure turbines are working, and three more low-pressure turbines have been installed. Other New Zealand fields, Ohaaki, Rotokawa, Mokai, Kawerau, Ngatamariki, Tauhara, and Ngawha, have been developed and between them have an installed capacity of 15,854 MWe.

4.1.6.2 Organic Rankine cycles

One of the technologies that has gained the most interest in recent years are ORCs [53], systems capable of generating electricity from low-temperature energy sources (less than 180°C), using working fluids whose evaporation temperatures are lower than that of water.

The main systems that make up an ORC are shown in **Figure 2**. To extract the geothermal resource, a pump is required, which is responsible for making the fluid reach the heat exchanger (evaporator), to later be returned to the geothermal reservoir or to be used in another process.

The path of the working fluid begins in a storage tank, from where it is pumped (normally with a centrifugal pump) to the evaporator, where the heat transfer from the geothermal resource to the organic fluid will take place. Once the desired temperature is reached, the fluid will pass to the axial turbine, to rotate its blades and thus obtain electricity through an electric generator. Finally, the fluid is sent to the second heat exchanger (condenser), where the temperature of the working fluid is lowered using cooling water. At the end of the cycle, the fluid returns to the storage tank and the process is repeated.

The cooling water is sent to the condenser to obtain the heat from the organic fluid, so it leaves the exchanger with a higher temperature. Therefore, a cooling tower is required to lower the temperature using fans and a pump to send the fluid back to the condenser. It should be noted that this step is omitted if the heat exchanger is replaced by an air condenser, since the working fluid will be cooled directly with air, so cooling water would not be required. This equipment is generally used when there is no water available at the installation site.

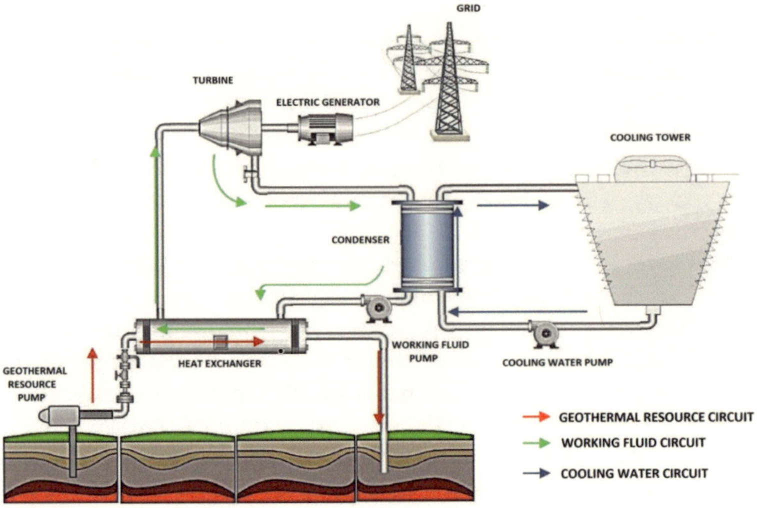

Figure 2.
Schematic diagram of the ORC diagram [by the authors].

Regarding the working fluids, these can be selected from a long list of candidates, including hydrocarbons, hydrofluorocarbons, siloxanes, and mixtures of these components [54], each with different thermodynamic properties.

Among the first commercial ORC-type plants were the following (Bronicki, 2017):

1. In 1952 in Kiabukawa, Congo, a small 200 kW unit was installed that used water at 91°C.
2. 1966, Ormat (Mali, Africa), creates a 0.6 kW solar turbogenerator, using dichlorobenzene.
3. 1967 in Russia 500 kW geothermal plant using R12.
4. 1979 at Kawasaki, Japan with 2900 kW using Freon 11.
5. 1979, Ormat built a 150-kW solar pond.
6. 1979 McCabe (California) builds a 12.5 MW ORC in cascade with two fluids, isopentane, and isobutane.
7. 1982, Ormat installed 15 kW in Mexico using freon 113.

5.Methodology for projects of direct use of geothermal energy, sustainable development

The implementation of direct use projects in Latin America has had an incipient growth with respect to other places in the world, however, it should be noted that with respect to electricity generation, there is a greater area of opportunity [55–57]. This methodology is intended to develop projects for direct use that generate a positive impact on society, the economy, and the environment. This seeks the understanding

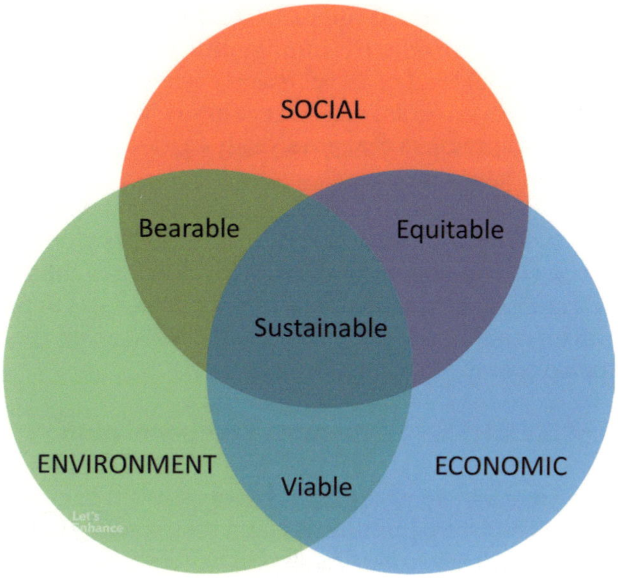

Figure 3.
Venn diagram of sustainable development at the confluence of the three pillars that characterize it.

and acceptance of these projects by the communities, which would serve as a spearhead for the development of larger projects, for example, electricity generation.

Sustainable development can be defined as a dynamic process, or an action plan or road map, toward a desirable future state for human societies in which living conditions and resource use continue to meet human needs without undermining the integrity, stability, and beauty of natural biotic systems. The efficient use of resources through saving and reuse provides an opportunity for each human being to develop freely, in balance with society and in harmony with the environment; that is, to avoid the loss, change, deterioration, impairment, adverse effect, or modification of the habitat, ecosystems, elements, and natural resources, of their chemical, physical, or biological conditions, of the interaction relationships that exist between them, as well as the environmental services they provide [58].

Therefore, it is considered that sustainable development is built on three fundamental pillars, which work in harmony for the gestation of true sustainable development, all with the aim of guaranteeing the right of every person to live in a healthy environment for their development, health, and wellness, see **Figure 3**.

5.1 Characterization of geothermal zones of low enthalpy

Currently, there are many works to estimate geothermal potential, where medium and low enthalpy resources stand out. This information is essential for the spatial location of the future project, information that will be concatenated with another set of data on its viability and type of project.

5.2 Hierarchy of areas with the highest probability of success

The hierarchization of zones of greater probability is a job that requires secondary sources, to carry out the analysis of the social, environmental, legal, and productive conditions of the geothermal zones of interest. This ranges from analyzing the number of men and women in the region, the immigration rate, and the distribution of the population in the educational level of 15 years and over. It is important to verify if the area of interest is not within an ecological reserve or one with a high environmental impact. The activities and products that are already carried out in the area to support them instead of inserting new ones.

With this analysis, it is possible to prioritize the areas that do not have environmental barriers, that have agro-industrial activities with thermal processes below 150°C and with a target population that wants to participate in the projects, with which a preliminary study is carried out feasibility by zones and in this way the target population is identified, that is, the one that has the resource and the area of opportunity for the development of direct use projects.

5.3 Measurement of the interest-acceptance of the population

This section of the methodology consists of generating instruments and tools that allow working directly with the communities, and these tools focus on the exploratory march and social mapping. These visual tools will help to engage the community with the project and it will be possible to detect the way in which the inhabitants perceive their space.

5.4 Choose lines of business model

The possibilities of successful productive projects are wide; however, projects that are understandable and accepted by the surrounding community are recommended, so the activities already carried out before the project must be taken into account so that their introduction is more natural [59].

Once said analysis is completed, the proposal will be presented to government entities or private investors in order to obtain the economic resources that contribute to the execution of the project.

5.5 Realization of the project

Finally, the project is developed, and in many cases, it is intended to be a turnkey project, which means that a 100% functional and operational production process is delivered to the community/client/investors, which implies the training of the personnel that plant or unit will be in operation. The steps to follow are grouped into four sequential steps and their acronym is called IDEA Development, an acronym that defines Identify, Develop, Evaluate, and Advance[2].

6. Conclusion

The use and efficient use of energy is a very popular topic in recent years to date; due to the energy crisis, the exhaustion of the main oil reserves, the growing demand for energy in the world, the reforms to the law in countries like Mexico, which speak of the right to a healthy environment, and last but not least, climate change.

[2] The IDEA methodology was developed by the iiDEA Applied Research Group, of the Engineering Institute of the National Autonomous University of Mexico.

Most DUs use geothermal energy left over from a previous process, this is known as cogeneration, and local geothermal development has many benefits, including social ones. The comprehensive use of resources accelerates the rates of return on investments, lowering the cost of energy, and consequently also increasing profits from the development of new comprehensive projects. However, the generation of jobs is the key factor that allows the integration of communities in the development of geothermal energy, leaving the doors open to the development of large projects such as geothermal power plants.

It is worth mentioning that most of the projects that have an agri-food purpose are aligned with the 17 sustainable development goals published by the UN, of which goals 2) Zero Hunger, 5) Gender Equality, 7) Affordable and non-polluting energy, 8) Decent work and economic growth, and 13) Climate action [58].

On the other hand, the production of electrical energy with binary cycles offers a great opportunity in the field of energy efficiency by being able to take advantage of the residual heat of industrial processes such as medium and low enthalpy geothermal energy to produce electricity, and recently, in applications of microsystems for the generation of heat, cold, and electricity (in so-called trigeneration applications) in homes or small commercial units, from the point of view of smart networks, which although they are in the demonstration phase, the investment and maintenance costs they are perceived as affordable and capable of saving primary energy with low GHG emissions [60].

Considering complementing the stationary electricity generation scheme with an on-site generation scheme, also known as distributed generation, through technologies such as binary cycles, will allow, among other aspects, to make the electricity supply more efficient, reduce transmission problems, decongest the electrical systems of each country, increase the efficiency of industrial processes through cogeneration schemes, thus reducing internal electricity demand, and as users, become independent from electricity companies, with economic benefits from the sale of electricity in those cases where there is great recovery potential.

Distributed generation is a relatively new concept that has been developed to reduce the operational problems and generation costs of electricity generation and transmission systems in a country.

The main characteristics of distributed generation are the following: 1) it reduces losses in the network by reducing energy flows to remote consumption areas, 2) the energy generated normally goes to the consumption centers and does not reverse flows in the transmission lines, 3) generation capacity generally ranges from a few tens of kW to 10 MW, and 4) for rural areas, distributed generation sources are generally mini or micro-hydroelectric, geothermal, and/or cogeneration plants that take advantage of waste from industries or agricultural to generate electricity on a small scale.

Some recommendations to encourage the development of electricity generation projects with low enthalpy resources are: 1) carry out a reassessment of the potential of the country/site of this resource with modern remote sensing technologies, supported by terrestrial measurements, 2) develop and adapt technology to quantify in detail the punctual resources (assessment of small shallow reservoirs), 3) develop and adapt technology to drill small geothermal wells, 4) technology to pump very hot water, but at a shallow depth, 5) develop or adapt the technology for small generation plants (<1 MW), and 6) include small geothermal energy in the legislation for the promotion of renewable energies.

Finally, to promote the development of small low enthalpy geothermal fields, it is required: 1) a good understanding of the size of the resource (how much hot water can be extracted without depleting it), 2) technology to extract hot water (drilling with preventers; extraction with deep well pumps), 3) economical and reliable technology for generating electricity (Turbines and associated equipment), and 4) legislation to market the energy generated (own uses, connection to the network).

Appendices and nomenclature

UNAM	Universidad Nacional Autónoma de México
EGS	Enhance Geothermal Systems
ORC	Organic Rankine Cycle
DUs	Direct Uses
CU	Cascade Uses
TOP	Tons of Oil Equivalent
GHP	Geothermal Heat Pumps

Author details

Héctor Aviña-Jiménez*, Eduardo Pérez-González and Rodrigo Alarcón-Flores
Institute of Engineering of the National Autonomous University of Mexico, Mexico City, Mexico

*Address all correspondence to: havinaj@iingen.unam.mx

References

[1] Gazo F, Lind L. Low Enthalpy Geothermal Energy - Technology Review [Repor 2010/20]. GNS Science; 2010

[2] Mburu M, Kinyanjui S. Cascade Use of Geothermal Energy: Eburru Case Study. Geo-Heat Center; Quarterly Bulletin; 2012. pp. 133-145

[3] Lund J, Boyd T. Direct utilization of geothermal energy 2015 worldwide review. In: Proceedings World Geothermal Congress 2015, Melbourne, Australia. 2015

[4] Van Nguyen M, Arason S, Gissurarson M, Pálsson P. Use of Geothermal Energy in Food and Agriculture - Opportunities for Developing Countries. Rome, Italy: Food and Agriculture Organization of the United Nations (FAO); 2015

[5] IRENA. Geothermal Power: Technology Brief. Abu Dhabí: International Renewable Energy Agency; 2017

[6] SENER. Uso Directo del Calor Geotérmico. MExico City: Secretary of Energy, Government of Mexico; 2018

[7] Christopher H, Armstead H. Energía Geotérmica. Mexico City: LIMUSA; 1989

[8] UNESCO. Geothermal Energy, a Review of Research and Development. París: Earth Sciences; 1973

[9] IGA, International Geothermal Association, 2010-2014. Available from: https://www.geothermal-energy.org/explore/our-databases/geothermal-power-database/#direct-uses-by-purpose. [Accessed: May 4, 2020]

[10] Huttrer GW. Geothermal Power Generation in the World 2015-2020 Update Report. In: Proceedings World Geothermal Congress 2020, Reykjavik, Iceland, April 26–May 2, 2020.

[11] Lund J, Toth A. Direct Utilization of Geothermal Energy 2020 Worldwide Review, Reykjavik: Proceedings World Geothermal Congress 2020, 2020

[12] Czellecz B, Petrea D. Mineral water for treatments: Summarized presentation of the nathing culture. In: Studia UBB Geographia, LVIII. 2013

[13] Lund J. Taking the Waters. Introduction to Balneology. Vol. september. Geo-Heat Center; 2000. pp. 2-8

[14] Lund J, Freeston D, Boyd T. Direct utilization of geothermal energy 2010 worldwide review. In: Proceeding World Geothermal Congress 2010. Bali, Indonesia; 2010

[15] Ana Q, Wanga Y, Jun Z, Chao L, Yan W. Direct utilization status and power generation potential of direct utilization status and power generation potential of Tianjin, China: A review. Geothermics. 2016;**64**:426-438

[16] B. X. H. C. a. L. D, Zheng X. Geothermal direct use and its contribution to CO2 emission saving in China. In: Proceedings World Geothermal Congress 2010; Bali, Indonesia, 25-29 April 2010. 2010

[17] Adele M, Davide S, Gabriele C, Eleonora B, Maurizio C, Cerutti PCP, et al. Geothermal energy use, country Update for Italy. In: European Gothermal Congress 2019, Den Haag, the Nertherlands, 11-14 June 2019. 2019

[18] Carella R, Sommaruga C. Italian agricultural uses of geothermal energy. Bulletin d'Hydrogrologie. 1999;(17)

[19] Smith KC. A Layman's Guide to GEOTHERMAL Aquaculture. Geo-Heat Center; 1981. p. 1981

[20] Lund J. Agriculture and aquaculture applications of geothermal energy. Geothermics. 1986;**15**(4):415-420

[21] Lund J. "Gone Fishing" Aquaculture Project Klamath Falls, Oregon. GHC Bulletin; 2003

[22] Johnson WC, Smith KC. Use of geothermal energy for aquaculture purposes. In: Phase iii - Final Report. Pacific Northwest Regional Commission; 1981

[23] Johnson WC. Culture of Freshwater Prawns (Macrobrachium Rosenbergii) Using Geothermal Waste Water. Geo-Heat Center; 1978

[24] Culver G. Generation, Fish Rearing Ponds Cascaded from Binary Power. Geo-Heat Center; 2005

[25] Árni R. Geothermal energy in aquaculture. In: Short Course VI on Utilization of Low- and Medium-Enthalpy Geothermal Resources and Financial Aspects of Utilization. El Salvador; 2014

[26] Georgsson LS, Fridleifsson G. High Technology in Geothermal Fish Farming at Silfurstjarnan Ltd., NE-Iceland. GHC Bulletin; 1996

[27] Jorquera C. Piensa en Geotermia. 2021. Available from: https://www.piensageotermia.com/operaciones-de-cultivo-de-salmon-a-gran-escala-utilizaran-la-energia-geotermica/?utm_source=rss&utm_medium=rss&utm_campaign=operaciones-de-cultivo-de-salmon-a-gran-escala-utilizaran-la-energia-geotermica&utm_source=Lista+de [Accessed May 15, 2020]

[28] Appelbaum S. Aquaculture experiences in the Negev Desert in Israel. In: FAO Technical Workshop, 6-9 July, 2011

[29] GREITZER Y, LEVITTE D. Geothermal Update Report from Israel. Israel, s.f;

[30] MacGibbon DJ. The Effects of different water quality parameters on prawn (Macrobrachium Rosenbergii) yield, Phytoplankton Abundance and Phytoplankton diversity at New Zealand Prawns Limited, Wairakei, New Zealand [Thesis]. Victoria University of Wellington; 2008

[31] F. A. Geothermal Activities in Algeria. In: Proceedings World Geothermal Congress 2010. Bali, Indonesia, 25-29 April, 2010

[32] Ouali S, Benaïssa Z, Belhamel M, Khellafa A, Kamel Baddari MD. Exploitation of albian geothermal water in South Algeria. Energy Procedia. 2011;**6**:101-109

[33] A. O. A., A. A., M. D., I. A., O. S., B. K., I. K, Updating of the most important Algerian geothermal provinces. In: Proceedings World Geothermal Congress 2020. Reykjavik, Iceland, April 26–May 2, 2020

[34] World Energy Council. World Energy Resources. England: World Energy Council; 2013

[35] European Geothermal Energy Council. Geothermal Utilization for Industrial Processes. EGEC, s.f

[36] Yasukawa K, Nishikawa N, Sasada M, Okumura T. Country Update of Japan. In: Proceedings World Geothermal Congress 2020. Reykjavik, Iceland; 2020

[37] Chiodi Agostina L, Filipovich RE, Esteban C, Pesce AH, Stefanini VA. Geothermal country Update of Argentina: 2015-2020. In: Proceedings World Geothermal Congress 2020, Reykjavik, Iceland. 2020

[38] Rojas F. Piensa en Geotermia. 2015. Available from: https://www.piensageotermia.com/

islandia-platanos-e-invernaderos-geotermicos/ [Accessed: June 22, 2020]

[39] Jorquera C. Piensa en Geotermia. 2018. Available from: https://www.piensageotermia.com/el-banco-holandes-rabobank-considera-que-la-geotermia-es-un-elemento-clave-para-las-operaciones-de-invernadero/ [Accessed: May 17, 2020]

[40] Rojas F. Piensa en geotermia. 2015. Available from: https://www.piensageotermia.com/de-larderello-a-antofagasta-la-relacion-geotermica-entre-enel-y-chile/ [Accessed: May 13, 2020]

[41] Rojas F. Piensa en geotermia. 2016. Available from: https://www.piensageotermia.com/enel-da-a-todas-sus-plantas-de-geotermia-en-la-toscana-sistemas-de-reduccion-de-emisiones/ [Accessed: May 13, 2020]

[42] Ormad A. Piensa en geotermia. 2014. Available from: https://www.piensageotermia.com/vapori-di-birra-cerveza-100-geotermica-italia/ [Accessed: May 13, 2020]

[43] Jorquera C. Piensa en geotermia. 2018. Available from: https://www.piensageotermia.com/nuevo-balneario-geotermico-se-esta-construyendo-en-un-lago-de-islandia/ [Accessed: June 9, 2020]

[44] Rojas F. Piensa en geotermia. 2015. Available from: https://www.piensageotermia.com/video-reportaje-de-la-apertura-del-proyecto-piloto-de-uso-directo-gdc-kenia/ [Accessed: June 10, 2020]

[45] Jorquera C. Piensa en geotermia. 2020. Available from: https://www.piensageotermia.com/gdc-y-condado-local-desarrollaran-conjuntamente-un-parque-industrial-geotermico-en-nakuru-kenia/ [Accessed: June 10, 2020]

[46] Richter A. Think Geoenergy. 2017. Available from: https://www.thinkgeoenergy.com/successful-maori-owned-dairy-operations-utilising-geothermal-new-zealand/ [Accessed: June 11, 2020]

[47] Jorquera C. Piensa en geotermia. 2018. Available from: https://www.piensageotermia.com/empresa-de-alimentos-japonesa-invierte-en-una-planta-de-procesamiento-de-lacteos-geotermicos-en-nueva-zelanda/ [Accessed: June 11, 2020]

[48] Jorquera C. Piensa en geotermia. 2019. Available from: https://www.piensageotermia.com/la-segunda-planta-lechera-geotermica-de-nueva-zelanda-se-encuentra-pronta-a-ser-terminada/ [Accessed: June 11, 2020]

[49] Jorquera C. Piensa en geotermia. 2018. Available from: https://www.piensageotermia.com/nuevo-parque-industrial-cerca-de-la-planta-geotermica-ngawha-nueva-zelanda/ [Accessed: June 11, 2020]

[50] Jorquera C. Piensa en geotermia. 2019. Available from: https://www.piensageotermia.com/ntga-aumenta-el-suministro-de-calor-geotermico-de-proceso-a-los-socios-industriales-de-kawerau/ [Accessed: June 11, 2020]

[51] rnz.nz. Miraka's on the Moove [Interview]. RNZ. Available from: https://www.rnz.co.nz/audio/player?audio_id=201859581; 2017

[52] Chiasson A. From Creamery to Brewery with Geothermal Energy: Klamath Basin Brewing Company. Geo-Heat Center; 2006

[53] J. Besong. State of art on ORC applications for waste heat recovery and micro-cogeneration for installations up to 100kWe. In: 70th Conference of the ATI Engineering Association. ELSEVIER;

Energy Procedia, no. 82, pp. 994-1001, 2015

[54] Artiere TAM. A world overview of the organic Rankine cycle market. In: IV International Seminar on ORC Power Systems. Milano, Italy; 2017

[55] E. Iglesias, R. Torres, I. Martínez-Estrella and Reyes-Picasso. Summary of the 2014 Assessment of Medium- to Low-Temperature Mexican Geothermal Resources. In: Proceedings World Geothermal Congress 2015. 2015

[56] P., C. M. F. Bona. Valoración y gobernanza de los proyectos geotérmicos en América del Sur. In; Comisión Económica para América Latina y el Caribe," CEPAL, Chile, 2016

[57] Tsagarakis KP. Shallow geothermal energy under the microscope: Social, economic, and institutional aspects. Renewable Energy. 2019

[58] Mondal P, Dalai AK. Sustainable Utilization of Natural Resources. Taylor & Francis Group; 2017

[59] Yasukawa K, Kubota H, Soma N, Noda T. Integration of natural and social environment in the implementation of geothermal projects. Geothermics. 2018;**72**:111-123

[60] Macchi E, Astolfi M. Organic Rankine Cycle (ORC) Power Systems Technologies and Applications. Vol. 107. Woodhead Publishing Series in Energy; 2017